SOUTH PACIFIC
AIR WAR

VOLUME 4

Buna & Milne Bay
June–September 1942

MICHAEL CLARINGBOULD
PETER INGMAN

AVONMORE BOOKS

South Pacific Air War
Volume 4: Buna & Milne Bay
June–September 1942

Michael John Claringbould
Peter Ingman

ISBN: 978-0-6486659-7-7

First published 2020 by Avonmore Books

Avonmore Books
PO Box 217
Kent Town
South Australia 5071
Australia

Phone: (61 8) 8431 9780
www.avonmorebooks.com.au

NATIONAL
LIBRARY
OF AUSTRALIA

A catalogue record for this
book is available from the
National Library of Australia

Cover design & layout by Diane Bricknell

Cover artwork captions:

Front: An 80th FS P-400 struggles to climb over Port Moresby in July 1942 with its well-worn engine. These Airacobras, originally built for the Royal Air Force, wore a European camouflage scheme. This one carries the markings of previous squadrons as most New Guinea P-400s changed hands at least once during their lengthy combat careers.

Back: Buntaciho Lieutenant Sasai Junichi flying Zero V-121 attacks a 3rd BG B-25C over the Huon Gulf as it flees after attack on Lae in June 1942.

CONTENTS

Michael Claringbould meeting the Cambodian Prime Minister Hun Sen, Phnom Penh, 2013.

Peter Ingman signing books at the WWII Weekend and Airshow, Reading Regional Airport, Pennsylvania, June 2019.

Michael Claringbould – Author & Illustrator

Michael spent his formative years in Papua New Guinea in the 1960s, during which he became fascinated by the many WWII aircraft wrecks which still lie around the country. Michael has served widely overseas as an Australian diplomat, including in South East Asia and throughout the South Pacific where he had the fortune to return to Papua New Guinea for three years commencing in 2003.

Michael has authored and illustrated various books on Pacific War aviation. His history of the Tainan Naval Air Group in New Guinea, *Eagles of the Southern Sky*, received worldwide acclaim as the first English-language history of a Japanese fighter unit, and was translated into Japanese. An executive member of Pacific Air War History Associates, Michael holds a pilot license and PG4 paraglider rating. He continues to develop his skills as a digital 3D aviation artist, using 3DS MAX, Vray and Photoshop to attain markings accuracy.

Peter Ingman – Author

A keen aviation historian, Peter's key interest is in the early stages of the Pacific War. Two of his books, *Zero Hour in Broome* and *Carrier Attack Darwin 1942* (both with Dr Tom Lewis) have received many favourable reviews in Australian media. A former business executive, Peter has travelled widely throughout northern Australia and the South Pacific conducting research for his books. He is the Chairman of the South Australian Aviation Museum History Group.

Other Books by the Authors
Michael Claringbould and Peter Ingman

South Pacific Air War Volume 1: The Fall of Rabaul December 1941–March 1942 (2017).

South Pacific Air War Volume 2: The Struggle for Moresby March–April 1942 (2018).

South Pacific Air War Volume 3: Coral Sea & Aftermath May-June 1942 (2019).

Michael Claringbould

Black Sunday (2000).

Eagles of the Southern Skies (2012, with Luca Ruffato).

P-39 / P-400 Airacobra versus A6M2/3 Zero-sen New Guinea 1942 (Osprey, 2018).

Pacific Adversaries Volume One Japanese Army Air Force vs The Allies New Guinea 1942-1944 (2019).

P-47D Thunderbolt versus Ki-43 Hayabusa New Guinea 1943/44 (Osprey, 2020).

*Pacific Adversaries Volume Two
Imperial Japanese Navy vs The Allies New Guinea & the Solomons 1942-1944* (2020).

Operation I-Go Yamamoto's Last Offensive – New Guinea and the Solomons April 1943 (2020).

Pacific Profiles Volume One Japanese Army Fighters New Guinea & the Solomons 1942-1944 (2020).

*Pacific Adversaries Volume Three
Imperial Japanese Navy vs The Allies New Guinea & the Solomons 1942-1944* (2020)

*Pacific Profiles Volume Two
– Japanese Army Bomber & Other Units, New Guinea and the Solomons 1942-1944* (2020).

Pacific Profiles Volume Three – Allied Medium Bombers, A20 Series, Southwest Pacific 1942-1944 (2021).

Peter Ingman

Zero Hour in Broome (2010, with Dr Tom Lewis).

Carrier Attack Darwin 1942 (2013, with Dr Tom Lewis).

Mark Vc Spitfire versus A6M2 Zero-sen Darwin 1942 (Osprey, 2019).

P-40E Warhawk versus A6M2 Zero-sen: East Indies and Darwin 1942 (Osprey, 2020).

The RAAF in South Australia 1939-45 (2020).

INTRODUCTION

This fourth volume of the *South Pacific Air War* series chronicles air war in New Guinea for the eleven weeks from 19 June until 8 September 1942. It can be read alone or as a continuation of the first three volumes that spanned the first six months of the Pacific War, culminating in the Battle of the Coral Sea.

The earlier three volumes cover a broad area of conflict in the wider South Pacific, continued in this volume until 7 August when the American invasion of Tulagi and Guadalcanal took place. From that date fighting in the South Pacific divides into two distinct fronts: New Guinea and the Solomon Islands. Subsequent coverage of both fronts becomes too unwieldy in a single and highly detailed narrative, hence after 7 August all Solomon Islands operations are discontinued from this volume, including Allied units operating from neighbouring island groups such as New Caledonia, the New Hebrides and Fiji. However, regular general references are made to the Solomons fighting as it impacted on the availability of Japanese airpower at Rabaul.

The eleven weeks covered by this volume was a critical period during which the strategic situation in New Guinea grew vastly more complicated. As the Allies secretly developed a major new base at Milne Bay on the eastern tip of Papua, the Japanese landed at Buna and began an overland assault over the Owen Stanley mountains. Then, in late August, the Japanese launched a two-pronged amphibious campaign to capture Milne Bay. While there had been virtually no land fighting at all up until June, by September land forces were engaged in an epic and bloody struggle that would continue in New Guinea until the end of the war in 1945 with the Wewak land campaigns.

The land struggles in this volume were launched in an environment of Japanese naval supremacy, albeit one in which their airpower was initially much depleted by the evolving conflict in the Solomons. Interestingly, and unlike the periods covered in the previous three volumes, no aircraft carriers appeared in New Guinea waters during this June – September period. Instead the land-based air forces were depended upon by their respective commanders more than ever. These air forces would need all their enterprise and ingenuity to meet the ever-increasing demands of the new and fast-changing environment.

Never before has this campaign been chronicled in such detail, with Allied accounts matched against Japanese records for a truly factual account of the conflict.

Michael John Claringbould & Peter Ingman
Canberra
August 2020

Glossary and Abbreviations

(Japanese terms in italics)

AA	Anti-Aircraft
AIF	Australian Imperial Force
ANGAU	Australian New Guinea Administrative Unit
AOB	Advanced Operational Base (RAAF)
BG	Bombardment Group (USAAF)
BS	Bombardment Squadron (USAAF)
Buntai	Equivalent to a *chutai* but usually accompanied by administrative or established command status.
Buntaicho	Leader of a *buntai*
Chutai	Japanese aircraft formation normally comprised of nine aircraft.
Chutaicho	Flight leader of a *chutai*.
CN	Constructor's Number
CO	Commanding Officer
Flyer1c	Aviator First Class (IJN)
FG	Fighter Group (USAAF)
FS	Fighter Squadron (USAAF)
FCPO	Flying Chief Petty Officer (IJN)
FPO1c	Flying Petty Officer First Class (IJN)
FPO2c	Flying Petty Officer Second Class (IJN)
Hikocho	Administrative commander of a *Kokutai*, senior to the *Hikotaicho* (operational commander).
Hikotaicho	commander of a *kokutai*
Hokoku	Inscriptions (translating as "patriotic") which signified that an aircraft was donated by an individual, organisation or corporation. The donor's name appeared in the *kanji* subscript.
HMAS	His Majesty's Australian Ship
IJA	Imperial Japanese Army
IJN	Imperial Japanese Navy
Katakana	Phonetic characters used in written Japanese, usually used for geographic place names.
Kokutai	An IJN air group, consisting of between three and six *chutai*.
Ku	Abbreviation of *kokutai*
POW	Prisoner of War
PRS	Photo Reconnaissance Squadron (USAAF)
PSP	Pierced Steel Planking (Marston matting)
RAF	Royal Air Force
RAAF	Royal Australian Air Force
Sea1c	Seaman First Class (IJN)
Shotai	A tactical formation typically of three aircraft (although sometimes two or four aircraft).

Shotaicho	Flight leader of a shotai	**USN**	United States Navy
SNLF	Special Naval Landing Force (IJN)	**USS**	United States Ship
		VMF-	Prefix for USMC Fighter Squadron
SoPA	South Pacific Area		
SWPA	South West Pacific Area	**VMO-**	Prefix for USMC Observation squadron, unusually in the case of VMO-251 which actually operated fighters in the South Pacific.
US	United States		
USAAF	United States Army Air Force		
USMC	United States Marine Corps		

Explanatory Notes

Place names are, where possible, consistent with 1942 usage, and with local spelling conventions, for example Nadi rather than Nandi.

Lieutenant is generally used as a standard term rather than referring to First Lieutenant, Second Lieutenant and Lieutenant (junior grade).

Measurements are also consistent with 1942 usage: generally, miles and miles per hour are used; Altitude is given in feet, even though Japanese aviators used metres.

Japanese individuals have their surnames first, followed by first name, as per Japanese usage.

Allied code names are used for most Japanese aircraft because these names are widely recognised by readers. However, this is not strictly historically correct, as some of these names had not yet become standardised during the period in question.

Note that the appendices of aircraft profiles in the earlier volumes have been discontinued. Instead readers are directed to the new *Pacific Profiles* series launched by Avonmore Books in 2020. This series is a better long term format for aircraft profiles as it allows some more detailed explanation of markings in the context of unit histories.

Addendum to Volume 3:

Four aircraft losses in early June (a Wirraway and three Beauforts) were omitted from Volume 3 and have been added to the loss tables in Appendix 1 of this volume.

p.210 of Volume 3: in the Japanese cumulative loss table, an entry for 10 x unknown types is missing. This has been added to the cumulative loss table in Appendix 3 of this volume. Note the total of 132 Japanese losses in Volume 3 is correct and has not changed.

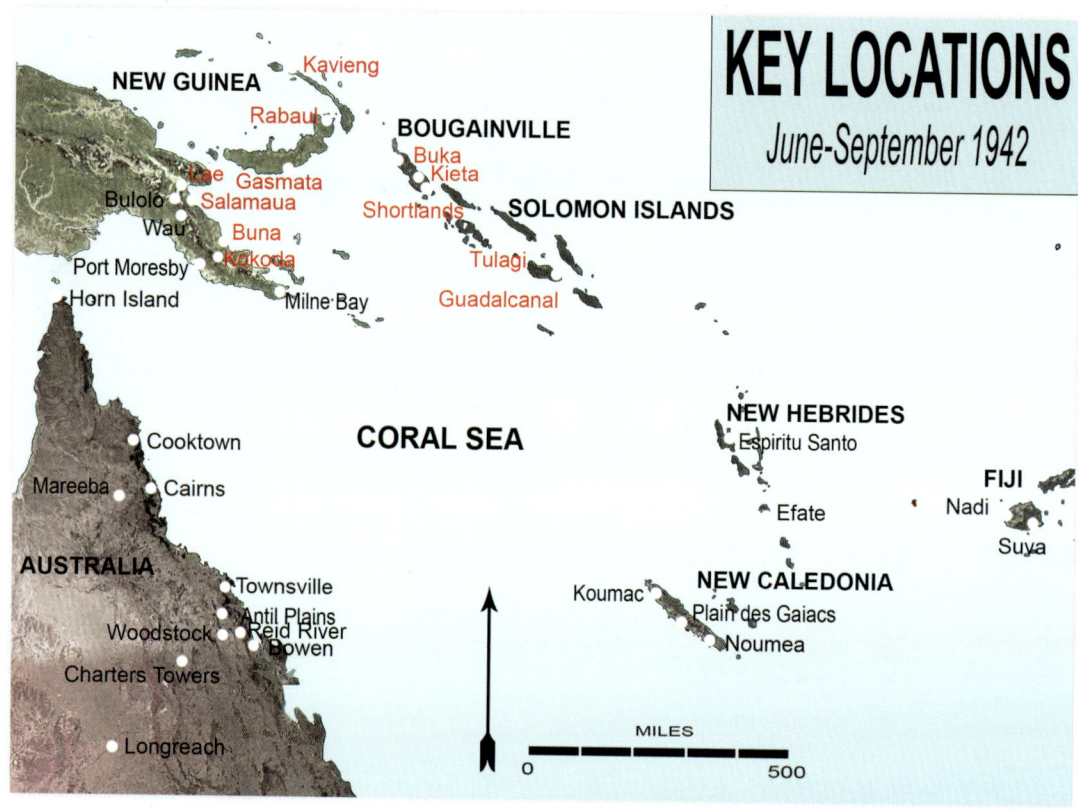

The wider South Pacific region as covered by this volume from 19 June – 7 August 1942. After the US landing in Guadalcanal on 7 August events in the Solomon Islands, New Hebrides, Fiji and New Caledonia are no longer covered as these areas became a separate theatre. Locations in red are those occupied by the Japanese, although the occupation of Buna and Kokoda did not occur until late July. Townsville and the cluster of airfields nearby were the home bases of most of the Allied Air Forces bomber units. In late July three squadrons of the B-17 equipped 19th BG moved from Longreach to Mareeba, which was considerably closer to New Guinea.

A map showing key New Guinea locations featured in this volume. Japanese occupied centres are in red, although Buna, Gona and Kokoda were not occupied until late July. Note that the area was divided into two territories, the Australian colony of Papua and Mandated New Guinea which was a former German colony administered by Australia after WWI. The distinction ceased to have any meaning after the outbreak of the Pacific War, although in the context of this title references to "Papua" usually mean the general area bounded by Port Moresby, Buna and Milne Bay.

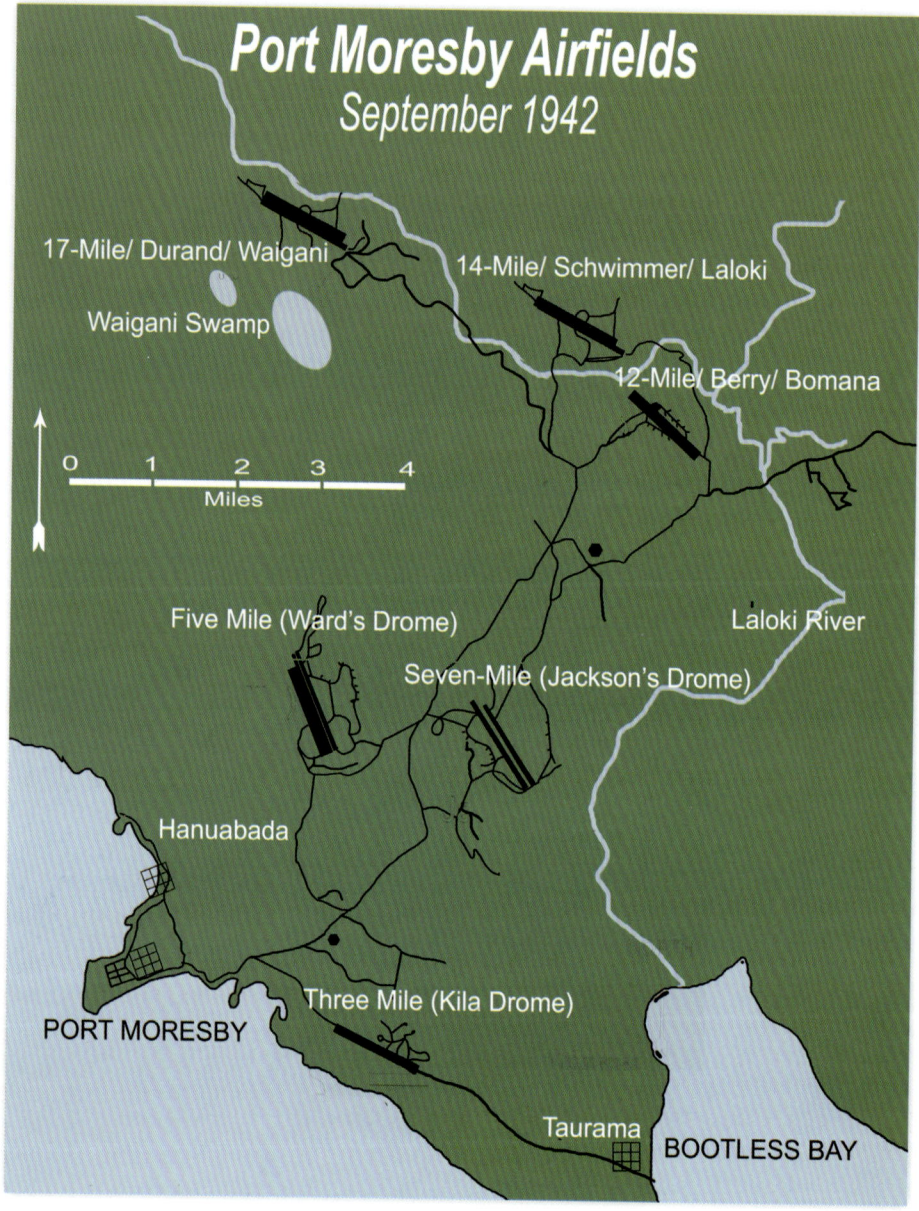

Port Moresby Airfields
September 1942

17-Mile/ Durand/ Waigani

14-Mile/ Schwimmer/ Laloki

Waigani Swamp

12-Mile/ Berry/ Bomana

0   1   2   3   4
Miles

Five Mile (Ward's Drome)

Seven-Mile (Jackson's Drome)

Laloki River

Hanuabada

Three Mile (Kila Drome)

PORT MORESBY

Taurama   BOOTLESS BAY

By September 1942 Port Moresby was a very busy aviation centre and was home to the six airfields seen in this map. In addition, another airfield had been built at 30-Mile (Rorona). The original pre-war civil 'drome was at Three-Mile (Kila) but was unsuitable for use by larger aircraft. Seven-Mile was the best facility and had been built by the RAAF in 1941. The other airfields had all been built in 1942 after an influx of engineering and construction units had arrived in Port Moresby and were in varying stages of development. Most of the airfields were eventually known by three names: their distance from Port Moresby central post office in miles, a geographic name and a name commemorating pilots and soldiers who had lost their lives in the area. Seven-Mile was known as Jackson's, after the CO of the RAAF's No. 75 Squadron, John Jackson, who was shot down and killed on 28 April. Today, the location still bears his name as Port Moresby's Jacksons International Airport. Berry, Durand and Schwimmer were all USAAF P-39 pilots killed between April and August. Ward's was named after an Australian army officer who helped build it but was killed in fighting on the Kokoda Track in August.

CHAPTER 1

OVERVIEW

The Allies

In June 1942 Australian and US airpower in New Guinea was organised under the Allied Air Forces command, led by Lieutenant-General George Brett, USAAF. Brett reported to General Douglas MacArthur, United States Army, who was the Commander-in-Chief of the South West Pacific Area (SWPA) command. The SWPA included all of Australia and New Guinea, while the Solomon Islands became part of the South Pacific Area (SoPA) which was under United States Naval control. Hence the fighting in Papua and Guadalcanal were controlled by different commands, further underlining the distinction between these two theatres.

Brett had at his disposal a reasonable but diverse collection of USAAF and RAAF units. The bulk of his airpower comprised several USAAF fighter and bomber groups which had arrived in Australia in early 1942. The fighter units were the 8th and 35th Fighter Groups equipped with P-39D/F Airacobras (another Fighter Group, the 49th, was in Darwin and is outside the scope of this volume). Two squadrons of the 8th FG had been based at Port Moresby since late April. From early June these were being replaced by two fresh squadrons from the 35th FG. All of these squadrons had been busy defending Port Moresby from Japanese raids and fighter sweeps which at times appeared daily. However, attrition had been extremely heavy, with some 59 Airacobras being lost by 18 June.

Brett's main striking power rested with three USAAF Bombardment Groups. The 3rd BG operated a mix of A-24 Banshees, Douglas A-20A Havocs and North American B-25C Mitchells. The 22nd BG was wholly equipped with Martin B-26 Marauders while the 19th BG operated Boeing B-17E Flying Fortresses. On paper this was a potent offensive force, but like the fighter groups these units had also sustained significant losses, both due to enemy action and capricious New Guinea weather conditions. Total losses by 18 June were 17 A-24s, 19 B-25s, 20 B-26s and 6 B-17s. Largely due to pressing needs in other theatres, replacements were generally not flowing in significant numbers to the SWPA at this time.

Other factors also seriously impeded the effectiveness of this force. As Port Moresby was too exposed to enemy attack, the bombers were based in north Queensland and flew up to Port Moresby for their sorties. Arriving late in the day or early in the morning the movements were timed to avoid the attention of Japanese raiders which generally appeared in the middle of the day. This arrangement meant there was usually a cumbersome delay of some 24-hours in launching any attack should a suitable target be detected. Overall, the use of Port Moresby as a forward base in this manner meant a great deal of flying for every mission launched against the enemy and correspondingly high wear on engines, airframes and aircrews.

A great deal of reconnaissance flying was also needed, much of which was drawn from these bombardment groups. Of the four B-17 squadrons that made up the 19th BG, the 435th Bombardment Squadron was mainly tasked with long range reconnaissance. Two squadrons of the 3rd BG operated B-25Cs, but during May these too had been heavily utilised for seawards reconnaissance, flying some 120 such missions.

In addition, maintenance facilities at north Queensland airfields were still rudimentary, with long delivery times needed to get spare parts delivered from southern depots. All of these factors meant that generally only half of the fighters and bombers on strength were serviceable at any one time. By early August, the USAAF could muster only 75 serviceable fighters (about half of which were in Darwin), 33 medium bombers and 43 heavy bombers. When allowing for reconnaissance needs and the draining factor of shuttling back and forth to Port Moresby for operations, a formation of half-a-dozen bombers of any type was around the typical striking force that could be assembled for most routine missions.

It should be noted that another USAAF heavy Bombardment Group, the 43rd, was in Australia at this time but was frustratingly without aircraft. The few new B-17s arriving across the Pacific had instead been used as replacements for the 19th BG. It was not until August that the 43rd BG was allocated its first Fortresses, and soon those began flying familiarisation missions under the guidance of the 19th BG. Also in August, B-25s belonging to the 38th BG began arriving in Australia, but these did not begin operations until after this volume ends on 8 September.

In addition, there was a single dedicated USAAF reconnaissance unit, the 8th Photo Reconnaissance Squadron. Equipped with unarmed Lockheed F-4 Lightnings, this squadron had commenced operations from Townsville in early May but after two accidents it would not resume operational duties until late July.

Alongside this USAAF force could be mustered several RAAF squadrons. No. 32 Squadron's Lockheed Hudsons were based at Horn Island and were largely employed flying anti-submarine patrols to protect the important supply convoys sailing between north Queensland and Port Moresby. Nos. 11 and 20 Squadrons, operating Consolidated Catalina flying boats, had recently withdrawn from Port Moresby to Bowen, south of Cairns. By this time just half-a-dozen Catalinas were on strength and the two squadrons were effectively operating as a single entity. The Catalinas' long range made them very useful, but they had proved too vulnerable to enemy fighters for most daylight reconnaissance missions. Instead their duties included night bombing missions against targets as far afield as Rabaul and Tulagi.

The RAAF placed much faith in its planned force of Bristol Beaufort torpedo bombers which were being manufactured locally in Australia. By mid-1942 the first squadron was operational. This was No. 100 Squadron, which had been named after and formed from the rump of a RAF torpedo bomber squadron (No. 100 Squadron, RAF) which had been largely annihilated while defending Malaya. On 22 May the unit had moved to north Queensland from New South Wales and commenced flying anti-submarine patrols from Mareeba. To help cover convoys, detachments of Beauforts operated from Port Moresby during June, and some of these aircraft would soon fly their first offensive operation from that location as will be described. However, this first experience of New Guinea operations for No. 100 Squadron would indeed be fleeting. To combat the Japanese submarine menace then threatening the shipping lanes off south-eastern Australia, the squadron was redeployed to Laverton in Victoria at short notice, and on 28 June the unit's fourteen Beauforts departed Mareeba for the long flight south.

Two RAAF fighter squadrons, both equipped with Curtiss P-40E Kittyhawks, were also based in Queensland. No. 76 Squadron was at Townsville where it had been undergoing operational training since April. Its sister unit, No. 75 Squadron, was also newly formed but had spent April as the sole defender of Port Moresby prior to being replaced by USAAF Airacobra squadrons. Losses had been heavy, with some twenty Kittyhawks and ten pilots lost. The squadron had been withdrawn to southern Queensland for rebuilding and would be recalled to New Guinea in late July.

No. 24 Squadron, flying Wirraways, was based at Townsville where it flew coastal anti-submarine patrols, but in July it would move to Sydney to re-equip with dive-bombers. A transport squadron, No. 33, was also based at Townsville and its Short Empire flying boats provided a regular transport service to Port Moresby. The sole permanent RAAF presence in New Guinea at this time was No. 1 Rescue and Communications Flight, operating a handful of aircraft including a DH.89 Dragon Rapide transport biplane and a DH.82 Tiger Moth. The unit's aircraft were light enough to fly into the many remote and often mountainous New Guinea airstrips from where they rescued downed Allied airmen and debilitated soldiers.

Of looming importance in the Papuan fighting was air transport, which had already been used to reinforce and resupply the Australian outpost in the mountains at Wau. However, military air transport capacity in the SWPA was slim. By mid-1942 just a handful of USAAF Douglas C-39s (a militarised DC-2) and Douglas C-53s (a paratrooper version of the DC-3, not as common as the ubiquitous C-47) had arrived in Australia by sea. These were supplemented by a collection of ex-Dutch transports which had escaped Java (Lockheed Lodestars, Douglas DC-2s, DC-3s and DC-5s) which had since been pressed into USAAF service. At times Australian civilian airliners were also available. The two USAAF transport units were the 21st and 22nd Troop Carrier Squadrons, operating from Archerfield (Brisbane) and Essendon (Melbourne) respectively.

As already outlined, the actual operational strength of the Allied Air Forces at this time was somewhat limited. In addition, there were serious problems with low morale in some American units in particular. These problems had surfaced during the Battle of the Coral Sea in May, when American bombers had failed to inflict a single casualty on the vulnerable Japanese Port Moresby invasion fleet. Also, confusion with ship recognition had led B-17s to attack friendly ships, fortunately ineffectively. Another factor affecting morale was aircrew exhaustion among the 19th BG, some of whom had been flying operations continuously since the Philippine and Netherlands East Indies campaigns in the first weeks of the war.

Against this background MacArthur lost faith in Brett and was particularly critical of the nature of the Allied Air Forces command which gave Australians control over key aspects of American operations. In many ways Brett was making the best of a bad situation, but there is evidence that MacArthur held a certain amount of personal antagonism towards him. Brett had in effect been MacArthur's predecessor until the appointment of the latter as the commander of the SWPA in April. Nevertheless, by early July MacArthur had communicated with the War Department in Washington about a replacement for Brett. As will be seen, the new replacement, Major-General George Kenney assumed command in early August. In time Kenney would provide excellent leadership and would institute far-reaching changes, including separating the American SWPA air units into a new organisation, the Fifth Air Force. However, most of these changes were not implemented until after the period for which this volume is concerned, during which time the Allied Air Forces operated under the far from ideal circumstances described.

In contrast to the airpower situation, the strength of Allied land forces in New Guinea had been growing steadily since the start of the year. This was aided by the peculiar fact that despite six months of warfare, there had only been one day of significant land fighting and that was back in January during the Japanese occupation of Rabaul. Even after March when the Japanese developed bases at Lae and Salamaua on the New Guinea mainland, there was no uptick in land fighting as these locations were maintained by naval base troops who did not penetrate far beyond these centres.

In late May, the Australian 14th Infantry Brigade had arrived in Port Moresby, doubling the size of the infantry available to defend the garrison. In addition, during the first half of 1942 a steady stream of service troops had been arriving. These included anti-aircraft batteries, engineering units and other support units of many different varieties. Accordingly, during this period Port Moresby had grown five airfields within its immediate vicinity. These enabled wide dispersal of the aircraft present to effectively ensure protection from all but the unluckiest of air raids. The main airfield with the best facilities was that at Seven-Mile, which was used by visiting bombers. The Airacobra squadrons occupied their own airfields which gave the fighters freedom to scramble when enemy aircraft were detected. These scrambles were aided by radar, which could not give cover over the mountainous spine of New Guinea, and a system of spotter outposts which usually provided warning of incoming raids. Without a timely warning the Airacobras were unable to reach the altitude of the enemy bomber formations.

Meanwhile the 7th Division, Australian Imperial Force, had arrived in Australia from the Middle East in March. This division comprised experienced and well-trained veteran soldiers seen as far more capable than the two militia brigades then in Port Moresby. After a period of rest and training it was decided to deploy the AIF troops to New Guinea. These would be used to reinforce Port Moresby and also to defend a new base to be established at Milne Bay on the far eastern tip of Papua. A survey party was sent to Milne Bay in early June and by the end of the month American engineer units had begun preparation of an airfield there. The site was on flat land that had been already cleared from the jungle for use as a coconut plantation. The construction work was undertaken under strict radio silence.

At this time small parties of Australian troops had been maintaining a close watch on Lae and Salamaua from nearby jungle hideouts. They were supplied via the mountain mining centre of Wau which had an airfield that had been established by airline companies in the 1930s to supply nearby goldfields. In late May, after the invasion threat to Port Moresby had passed, it was decided to fly in more troops to Wau. This was done, despite Wau being less than 50 miles by air from the Japanese airbase at Lae! While cloudy mountain conditions often hid the area from Lae, interception by Zeros was a constant danger for the transports flying between Port Moresby and Wau. Fortunately, Wau was close enough to Port Moresby to enable protective escort of the transports by Airacobras. However, no permanent air transports were then allocated to Port Moresby and future air supply of Wau would remain problematic until such a capability was provided.

By June the collection of troops supplied from Wau, known as Kanga Force, had grown to a force of several hundred men including a company of AIF commandos. It was decided to launch raids against Salamaua, where the garrison was estimated at just 250 Japanese, and Heath's Plantation, which was a small outpost protecting Lae. The raid on Salamaua went ahead in the early hours of 29 June and was a textbook commando operation. It was judged to be highly successful by the Australians who claimed to have killed 100 Japanese. While the actual count was only 18 Japanese casualties, captured equipment and documents proved to be very valuable for intelligence purposes. Among the Japanese killed was floatplane pilot Warrant Officer Nemoto Kumesako, whose captured diaries provided much insight into the early IJN operations in the South Pacific and which are often cited in the first three volumes of *South Pacific Air War*.

The raid on Heath's Plantation followed soon afterwards in conjunction with American air support. It was less successful than the Salamaua raid, and although resulting in ten Japanese casualties the main result of both of these operations was a frenzied Japanese response, both in the air and on the ground. Indeed, the situation in Papua was quickly evolving into something much more complex for both sides. Meanwhile, in addition to the Allied Milne Bay development, planning was underway for another new airfield on the north coast of Papua in the Buna area. On 11 July an RAAF Empire Flying Boat transported a party of American and Australian officers to inspect the area. However, they judged the coastal area to be unsuitable for an airfield due to poor drainage among other factors.

Allied Air Forces units, New Guinea & North Queensland, late June 1942

USAAF:

8th FG:

35th and 36th FS, P-39D/Fs & P-400 Airacobras, Woodstock & Townsville, Queensland (plus the 80th FS training in southern Queensland)

35th FG:

39th and 40th FS, P-39D/Fs & P-400 Airacobras, Port Moresby, Papua (plus the 41st FS training in Sydney)

3rd BG:

8th BS, A-24 Banshees, Charters Towers, Queensland

89th BS, A-20As, Charters Towers, Queensland (training)

13th and 90th BS, B-25C Mitchells, Charters Towers, Queensland

19th BG:

28th, 30th, 93rd and 435th BS, B-17E Flying Fortresses, Longreach & Townsville, Queensland

22nd BG:

2nd, 19th, 33rd and 408th BS, B-26 Marauders, Antil Plains, Reid River & Townsville, Queensland

Independent Squadron:

8th PRS   F-4 Lightnings, Townsville, Queensland

RAAF:

Nos. 11 and 20 Sqns, Catalinas, Bowen, Queensland

No. 24 Sqn, Wirraways, Townsville, Queensland

No. 32 Sqn, Hudsons, Horn Island, Queensland

No. 33 Sqn, Empire flying boats, Townsville, Queensland

(No. 75 Squadron, Kittyhawks, training in southern Queensland)

No. 76 Sqn, Kittyhawks, Aitkenvale Weir (Townsville), Queensland (training)

No. 100 Sqn, Beauforts, Mareeba, Queensland

No. 1 Rescue and Communications Flight, DH.82 & DH.89, Port Moresby, Papua

---

The Japanese

Significantly, a noteworthy moment during the planning of the South Pacific war occurred in the southern Japanese city of Fukuoka on evening of 7 June 1942, where the headquarters detachment of the newly-created 17th Army had just arrived from Tokyo after a five-hour flight. The party of senior officers was ordered to wait an extra night in their lodgings in order to receive an urgent briefing, provided by a staff officer the next night at 2100. The core of the briefing was that the Japanese Navy had suffered a setback at Midway and accordingly it had been decided to delay Operation FS, as outlined in detail below.

This key episode underlines how useful it is to understand the Japanese mindset and associated command structures of the times. These highlight two key differences to their Allied counterparts. First, Japanese objectives hinged around empire-building whereas Allied ones were simply to counter Japanese militarism. Second, Japanese objectives were constantly thwarted by interservice rivalry between the Army and Navy. As a result, plans were often delayed and complicated by trying to incorporate the competing, and often conflicting, needs of both services.

Back on 15 May 1942 an IJA / IJN central agreement had divided South Pacific operations into two key areas: New Guinea and the rest. Three days later the 17th Army was created to help achieve these goals, the first operational army to be established since the commencement of hostilities with the Allies. Its commander was Lieutenant-General Hyakutake Haruyoshi, who was ordered to cooperate fully with the IJN Fourth Fleet in securing South Pacific objectives.

However, during this period there was significant bickering between the Army and Navy over these plans. This bickering even extended to disagreement over the future administration of the captured territories. For example, the IJA was keenly interested in control of valuable nickel resources in New Caledonia. The end result of these machinations was a cadre of bureaucrats, politicians and military staff officers scurrying and arguing within Tokyo's corridors of power at a time when Allied strength was growing daily, and decisive action was badly needed.

After the Battle of the Coral Sea in early May, the seaborne invasion of Port Moresby had been postponed. However, the IJA was largely unconcerned, viewing the battle as a standard naval engagement, and one that had been won at a tactical level by the IJN. As such, in the worst case this was a major inconvenience.

Meanwhile the IJA drew up new plans, known as Operation FS, to seize Fiji, New Caledonia and Samoa as well as Port Moresby in July. However, as noted above, following the destruction of the main IJN carrier force at Midway in early June, Operation FS was postponed for two months and eventually cancelled. Despite this, another major outcome of these deliberations was to commence an immediate feasibility study of seizing Port Moresby via an overland route. This study was to be conducted in secret and included reconnaissance flights over northern Papua and the Kokoda area.

In preparation for Operation FS an IJA force had been assembled for the planned operations. The existing 5,000-strong South Seas Force in Rabaul was trebled in strength by adding other units, including an infantry regiment from the Philippines. Military Historian Peter Williams describes the new force as:

> ... best visualised as a light division with a stronger than usual proportion of engineering, labouring and medical support.

After much debate it was agreed to land an advance party of the South Seas Force in the Buna-Gona area to conduct a reconnaissance to study the land route to Port Moresby over the mountains via Kokoda. However soon after the force had been landed on 21 July the plan morphed into an overland offensive to capture Port Moresby and further troops soon followed.

Note up until this time all local naval operations in the South Pacific had been the responsibility of Vice-Admiral Inoue Shigeyoshi's Fourth Fleet, headquartered at Truk. However, there was a major IJN reorganisation after the losses at Midway in June. Accordingly, in mid-July the Eighth Fleet was formed, under Vice-Admiral Mikawa Gun'ichi, and this force assumed control of all South Pacific naval operations.

Mikawa had a sterling reputation within the IJN, and throughout the 1920s and 1930s had served, among other highly visible posts, as instructor at the IJN Torpedo School. He had been a key delegation member to the 1930 London Naval Treaty meetings and subsequently served as naval attaché in Paris, giving him further opportunity to showcase his cosmopolitan credentials. Later he held sea commands including the heavy cruisers *Aoba* and *Chokai* and the battleship *Kirishima*. Mikawa had been promoted to Rear-Admiral in 1936 and for the next two years served as chief of staff to the Second Fleet. Promoted to Vice-Admiral in 1940, Mikawa was competent, had a thorough knowledge of all aspects of IJN operations and was widely respected.

Mikawa arrived at Rabaul at the end of July with his 103-man staff where he setup his headquarters. The forces allocated to the Eighth Fleet were similar in strength to those used by the Fourth Fleet, comprising several cruisers, four destroyers, two small submarines and the seaplane tender *Kiyokawa Maru*. The *Kiyokawa Maru* had returned to Rabaul in June after repairs in Japan following battle damage inflicted by USN aircraft during the landings at Lae in March. The ship was reunited with its air group which had meanwhile remained in the South Pacific, operating from shore bases.

Among other forces allocated to the Eighth Fleet were a miscellany of light forces: mainly submarine chasers, patrol boats and minesweepers. Also included were the naval ground troops of the 3rd Kure Special Naval Landing Force (SNLF), the 5th Sasebo SNLF and five Establishment Units. The Establishment Units each had strengths of over 1,000 men and were mainly made up of unarmed construction personnel with the purpose of building air bases. Two of these units were landed at Guadalcanal in July to construct an airfield there, and another was destined for Buna. Thus, aside from the overland advance to Port Moresby, it is an irony that Japanese plans at this time largely mirrored the Allied intentions of building new South Pacific airfields.

Japanese airpower at Rabaul comprised four naval air groups called *kokutai* which made up the 25th Air Flotilla. The main striking power rested with the No. 4 *Kokutai*, operating G4M Betty bombers. This had been supplemented in late April with the Genzan *Kokutai* with its older G3M Nell bombers, but by July this unit had returned to Japan. No. 4 *Ku* had seen much action and had lost 26 Bettys since first arriving in the theatre. By June just 13 were operational. These bombers were based at Vunakanau near Rabaul, which like the Port Moresby airfields had been greatly expanded. Many protective revetments had been built such that losses on the ground to Allied air raids were rare.

Fighters in the area were operated by the famed Tainan *Kokutai*, which boasted a cadre of extremely skilled and experienced pilots to fly its formidable Model 21 Zero fighters. However, this force had also suffered heavy attrition, with as few as 20 machines operational after the Coral Sea operations. Most of these were forward based at Lae with others providing for the air defence of Rabaul from the fighter strip at Lakunai. A few venerable A5M Claudes were still in service at Lakunai tasked primarily with patrol duties, although these dated open-cockpit fighters had been used to defend against Allied raids as recently as May. However, deliveries of replacement Zeros were fairly frequent, including 15 flown from Truk on 25 May. The Tainan *Ku* also operated a handful of C5M Babs reconnaissance aircraft, one of which had overflown Horn Island on 17 June.

The other air unit making up the 25th Air Flotilla was the Yokohama *Kokutai* which operated Kawanishi H6K Mavis flying boats. Like the RAAF Catalinas, the healthy endurance of the Mavises was put to good use in flying long-range reconnaissance missions, although they continued to prove vulnerable to USN carrier-based fighters aided by radar direction. By June just half a dozen remained, with a few forward deployed to Tulagi, although a few replacements soon bolstered numbers to ten. The Yokohama *Ku* had recently raised a fighter *chutai*, which arrived in Rabaul in early June with A6M2-N Rufe floatplane fighters. These too were soon sent to Tulagi.

While the 25th Air Flotilla had sustained significant attrition, at least the IJN command had the option of sending reinforcements to Rabaul which could be flown in relatively quickly from the Central Pacific or Japan itself. Further, a Japan-based formation, the 26th Air Flotilla, was earmarked for service at the new airfield being constructed on Guadalcanal, from where Betty raids could be undertaken against Efate in the New Hebrides and long range Kawanishi H8K Emily flying boats could raid New Caledonia. Following the surprise seizure of Guadalcanal by US Marines in early August, various *kokutai* including those of the 26th Air Flotilla remained available to reinforce the existing South Pacific air units. As will be explained, some of these new units flew in to the newly expanded and refurbished airbase at Kavieng on New Ireland.

Land-based IJN air units, New Guinea and the Solomons, late June 1942

Tainan *Kokutai*	Lae & Rabaul, A5M4 Claudes & A6M2 Zeros
No. 4 *Kokutai*	Rabaul, G4M1 Bettys
Genzan *Kokutai*	Rabaul, G3M2 Nells
Yokohama *Kokutai*	Rabaul & Tulagi, H6K4 Mavises & A6M2-N Rufes

Kiyokawa Maru air component (shore-based, widely dispersed): E13A1 Jakes, F1M2 Petes, E8N2 Daves, E7K Alf

---

Tactics and Strategy

During the first six months of war in the South Pacific a key determining factor of each step of the campaign was the appearance of carrier forces by either side. However, after the Battle of the Coral Sea in May carriers would not return to the New Guinea area for many months. The Japanese had lost four of six fleet carriers at Midway. Of the two others, by late June *Shokaku* was still undergoing repairs in Japan and *Zuikaku* was involved in the Aleutian operations in the North Pacific. By comparison the USN had four fleet carriers in the Pacific, but initially time was needed to work up replacement air groups after the Midway losses. Ultimately three carriers (USS *Enterprise*, *Saratoga* and *Wasp*) would participate in the Guadalcanal landings in August while the USS *Hornet* remained in reserve at Hawaii.

In the absence of carrier-based airpower, the opposing land-based air forces in New Guinea slugged it out. The B-17s of the 19th BG were theoretically the most powerful strike force in the theatre, but so far had failed to land any killer blows on the enemy. Bombing from high altitude was usually inaccurate and cloud often obscured targets entirely. Despite this the appearance of B-17s worried the Japanese greatly as to date they had failed to shoot one down in New Guinea despite many attempts, although unbeknownst to them one had been forced down in a swamp due to combat damage incurred over Rabaul . On the other hand, Japanese fighters had forced a small but significant victory: B-26 raids on Rabaul had been abandoned due to the heavy losses incurred. Indeed, many Allied raids would now take place at night which meant an almost nil chance of accurate bombing.

On occasion both sides had inflicted significant damage to opposition ships and aircraft by means of low-level strikes. However, for various reasons this remained a tactic that would not be fully exploited until 1943. While raids against airfields were routinely conducted by both sides, investments in protective revetments and dispersal areas, particularly at Vunakanau and Kavieng, made such raids increasingly ineffective unless aircraft were unlucky enough to be caught in the open.

On 17-18 June the Japanese had changed their tactics and had instead attacked Port Moresby harbour, sinking the *Macdhui*. Such targeting of supply ships potentially offered higher dividends than the targeting of airfields. Likewise, the Allied Air Forces tried to catch the Japanese ships bringing supplies to Lae. However, from July the air forces of both sides would be called on for a new task: direct air support for land forces. This was initially in the form of ground attacks and reconnaissance but air transport and the air-dropping of supplies would also soon be demanded. Hence the calls on limited land-based air power were ratcheted up, and a growing sense of urgency and desperation was brought to the theatre.

In the absence of aircraft carriers, both sides used their naval power conservatively and mainly to protect supply ships and troop transports. The exception was the use of submarines, which offered options for offensive missions. In late May several IJN ocean-going submarines had been involved in the midget submarine attack in Sydney harbour. Subsequently four of these vessels remained off the coast of New South Wales where they targeted coastal shipping and within a fortnight had sunk three ships. This was the start of a concerted IJN submarine offensive off the east coast of Australia which lasted until mid-1943. While the details of this campaign are beyond the scope of this volume, it should be noted that it diverted a significant portion of RAAF resources to maintain anti-submarine patrols over wide areas of the Australian coast. Evidence of this was the June 1942 withdrawal of RAAF No. 100 Squadron from north Queensland to Victoria as already noted.

While the supply convoys running between north Queensland and Port Moresby represented potentially juicy targets for submarines, the shallow and reef-strewn waters in this area were unsuitable for operations by ocean-going submarines. The Eighth Fleet had two small, short-range submarines on strength, the *RO-33* and *RO-34*, which were more suited for operations in shallow coastal waters. Both had already operated off Port Moresby during the Battle of the Coral Sea after which they returned to Japan for repairs and overhaul. During August both vessels would again venture into the Gulf of Papua with significant ramifications as will be seen.

Meanwhile, in April some eleven S-class USN submarines had been allocated to the SWPA and were based in the Brisbane. Despite being of small size and dating from the early 1920s, from May these vessels commenced patrols in New Guinea waters. Due to the volcanic nature of Rabaul and its environs, there was ample deep water adjacent to shipping lanes which offered good prospects for submarine operations. These submarines had already sunk two IJN vessels in this area, as described in *Volume 3*, and would soon further add to that tally.

Thus, despite much changing of plans in the light of extended and often convoluted IJA / IJN negotiations, July would see the Japanese occupy Buna on the north coast of Papua and begin pushing inland. A month later they landed at Milne Bay where the Allies had constructed their first major new base in New Guinea since the start of the war. These developments, together with the opening of a major land campaign high in the Owen Stanley mountains, saw a frenzied response by airpower from both sides. All of this took place against the background of the landmark and surprise American landing at Guadalcanal in early August which triggered a whirlwind of escalation and a bevy of new IJN air units committed to the South Pacific.

Overall, the ten-week period from late June until early September saw the biggest changes in the South Pacific air campaign since the start of the war. From these transformations, the IJN air forces based at Rabaul suddenly found themselves fighting on multiple fronts, in response to which they were forced to make reactive decisions, not proactive ones. The air war was about to change, and irrevocably so.

On 25 June 1942 Squadron Leader Charles Sage in Beaufort T9604 tries to get back through the Kokoda Pass to Port Moresby after attacking Salamaua.

CHAPTER 2

NIGHT RAIDS
REGIONAL OPERATIONS 19–30 JUNE

The third volume of *South Pacific Air War* ends on 18 June 1942 with the successful raid on Port Moresby harbour by the No. 4 *Kokutai* which sank the merchant ship *Macdhui*. Port Moresby would now enjoy a break of almost a week before daylight raids resumed.

At this time the 19th Bombardment Group had been preparing one of its "all-out efforts" against Rabaul. At dawn on 19 June an impressive thirteen Fortresses departed Seven-Mile (seven from the 30th BS and six from the 93rd BS), followed two hours later by three from the 28th BS. A severe storm soon broke up the formation, however, and most of the pilots elected to abort the mission. Several Fortresses persevered and reached the target area, where there was a brief skirmish with two Tainan *Ku* Zeros: during the exchange one B-17 and one Zero sustained light damage. The IJN cargo ship *Wayo Maru*, primarily used as a water carrier, received minor damage in Rabaul Harbour from air attack on this date.

An attempt by three B-17s from the 435th BS to reach Rabaul that afternoon was cancelled due to the bad weather. The trio instead bombed Lae the following morning of 20 June. Eight Zeros rose to intercept but couldn't catch the Fortresses.

Also targeting Rabaul were five RAAF Catalinas, led by Squadron Leader Francis Chapman, which departed Cairns to make a night attack on Rabaul. Two turned back due to the weather, but the other three were able to drop a mix of 250- and 500-pound bombs on warehouses near the wharf area. One fire was reported.

On 21 June, the 2,626-ton auxiliary gunboat *Keijo Maru* was en route to Tulagi carrying communications equipment for the Yokohama *Ku* base there. About ten miles from its destination the ship was torpedoed and sunk by the submarine *S-44* with the loss of 63 sailors. Survivors were rescued by the minesweeper *W-20*. On leaving the area the *S-44* reported being bombed by an aircraft which resulted in some minor damage. The aircraft was a Tulagi-based Yokohama *Ku* Mavis piloted by Warrant Officer Hamano Torao who reported that his attack produced "unknown results".

A regular mission for the 435th BS Fortresses was lone reconnaissance sorties over enemy bases including Lae, Kavieng and Rabaul. Zeros often tried to intercept these flights, usually without success as the B-17s were always flying fast and at extremely high altitude. On 23 June Rabaul received forewarning of such a sortie and at 1230 Lieutenant Sasai Jun'ichi led fourteen Zeros into the air to try and catch the intruder. The B-17 crew reported interception by five Zeros but sustained no damage. However, three of the crew were then affected by altitude sickness, which forced the pilot to descend to 18,000 feet. This enabled another Zero to fire on the Fortress, but the only damage was a bullet hole in the port elevator of the bomber.

Also on 23 June a DC-2 of the 21st Troop Carrier Squadron flown by Lieutenant Robert Gerling had just landed at Charters Towers when a cross wind swerved the airliner off the runway. It collided with a Ford truck operated by the Queensland Main Roads Commission, structurally damaging the airframe to the extent that it was a complete loss. Originally a civilian airliner operated by the Dutch in the Netherlands East Indies as PK-AFK, the aircraft had been transferred to the USAAF in March 1942 and reallocated the Australian callsign VHCXF.

The following day, 24 June, a successful reconnaissance mission over Rabaul sent back a detailed report. Among ten ships noted in the harbour was a seaplane tender (the *Kiyokawa Maru*, just arrived from Japan) along with eleven flying boats and sixteen floatplanes. Six Fighters and fourteen bombers were noted at Lakunai, while observations of Vunakanau included the note that:

> Hard used road to edge of coconut plantation suggests dispersal of aircraft. Heavy construction activity revealed by two new taxiways and dispersal area N.W. and S.E, of runway.

The day also saw long-range Japanese reconnaissance missions of the Australian mainland by two Tainan *Ku* C5Ms. Departing Lae that morning, around midday the aircraft were over their assigned areas of Cairns and Cooktown. Nothing of interest was noted, as no aircraft were based at these coastal locations, but the Japanese were correct that Allied aircraft were regularly operating in this area. They had probably detected increased radio traffic in the vicinity of Mareeba, just 25 miles inland from Cairns, which for the past month had been used by the Beauforts of No. 100 Squadron, RAAF.

Late on 24 June nine B-17s departed Seven-Mile for a night attack on Rabaul. Seven of the bombers reached the target area, and as well as bombing some 500 rounds of 0.50-inch calibre machine gun fire was expended in strafing searchlights. As a result of the bombing the crews reported fires visible from 60 miles away.

From 22 June three Beauforts had been temporarily operating from Port Moresby flying seawards reconnaissance and anti-submarine patrols. On the evening of 23 June they used their night-flying capability to provide training for searchlights operated by American anti-aircraft batteries. By day the Beauforts were helping to cover the first convoy taking troops and supplies from Port Moresby to Milne Bay. This comprised two Dutch merchant vessels escorted by the sloop HMAS *Warrego* and the corvette HMAS *Ballarat*. These ships arrived at Milne Bay on 25 June and had completed unloading three days later, in an operation undetected by the Japanese.

At this time a supply ship was detected making a run towards Lae. This was the 3,114-ton auxiliary gunboat *Seikai Maru*, and during such voyages the ship typically unloaded and departed under the cover of night. However, it was decided to make use of No. 100 Squadron's night training whose pilots were eager to mount their first operational mission. Accordingly, on 25 June five Beauforts were flown into Port Moresby from Mareeba, where they joined two others already present. All seven aircraft would take part in the bombing attack that evening.

After departing Seven-Mile at 2100 guided by a lit flare path, the Beauforts headed for Cape Ward Hunt on the northern Papuan coast from where they would split into two sections. Wing Commander JR "Sam" Balmer in A9-46 led five Beauforts which would hunt for shipping

An RAAF Beaufort seen at Wards 'drome in Port Moresby in 1942. The initial deployment of No. 100 Squadron Beauforts to New Guinea in June was brief, but the unit would return to New Guinea with its torpedo bombers in early September.

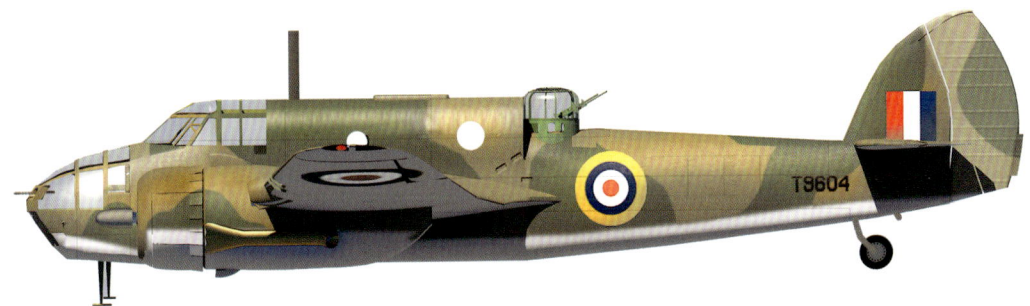

Bristol Beaufort T9604, No. 100 Squadron, RAAF. Flown by Squadron Leader Charles Sage and lost in New Guinea on 26 June 1942.

in Lae harbour, while Squadron Leader Charles Sage in A9-52 would lead the other two in a diversionary attack on Salamaua. It was thought that the ship in question was just then approaching Lae, and when it received news of the attack on Salamaua it would head back out to sea and hence present an easier target free of Lae's AA defences. The Beauforts were preceded by four 3rd Bombardment Group B-25s that bombed and strafed Salamaua from 400 to 800-feet and reported starting a fire.

The Beaufort pair led by Sage bombed and strafed Salamaua from 1,500 feet at 0945, and in a rare case of a plan working as intended, a short time later Balmer came upon the ship steaming 30 miles east of Lae in the moonlit waters of Huon Gulf. Having to make multiple low-level runs over the ship because his bomb release would not initially function, both Balmer and two other Beauforts believed they had hit the vessel. One of the Beauforts, A9-31 flown by Squadron Leader CS Bernard, was hit by AA fire from the ship. Suffering a loss of hydraulic power, Bernard subsequently made an emergency belly landing at Seven-Mile. Meanwhile, two other Beauforts dropped their bombs on AA positions at Lae.

The Beauforts encountered low cloud on their return flight, although six were back safely in the early hours of 26 June included the damaged A9-31. At 0344 Squadron Leader Sage was heard requesting a Radio Direction Finding bearing. The aircraft was estimated to be 60 miles to the north-west of Port Moresby, but Sage and his three crewmen vanished without a trace. The wreck of the Beaufort was not found until 1987, at an altitude of 7,000-feet in a mountainous area north-west of Port Moresby.

Balmer and his pilots were confident that all three Beauforts that attacked the ship had scored hits. The belief that they sunk the ship is underscored in the squadron log that records:

> Hits were scored by each aircraft and ship confirmed sunk by reconnaissance aircraft the following day.

However, there is no evidence of any vessel being damaged let alone sunk, and it appears that the crews misjudged bomb explosions for hits in the dark conditions (in fact *Seikai Maru* was later sunk off Kavieng on 16 September 1943). As explained earlier, within days of this mission all of the Beauforts had returned to Mareeba and then begun flights south to Laverton in Victoria where No. 100 Squadron was being relocated.

The theme of night operations at Port Moresby continued with a night raid by Mavis flying boats on 24 June. Yokohama *Ku Hikocho* Lieutenant-Commander Tashiro Soichi led a trio of Mavises from Rabaul harbour at 1540. That evening they dropped 36 x 60-kilogram bombs over Port Moresby before returning to base at midnight. Heavy low cloud was present with the result that the bombs fell into the savannah plains which lie several miles east of Seven-Mile. Two Airacobras aided by searchlights attempted to intercept the intruders but were unsuccessful.

The following evening Lieutenant Adachi Yoshiro departed for Port Moresby with a pair of Mavises on a similar mission, but due to bad weather (this was the same night that Sage's Beaufort was lost) they were forced to return to base when barely halfway to the target.

Earlier on this day, the 25 June, Lieutenant-Commander Nakajima Tadashi led 21 Tainan *Ku* Zeros on a fighter sweep against Port Moresby. This was the first such offensive mission in a week after the unit had been largely stood down for maintenance, and Nakajima's 21 fighters represented every serviceable machine at Lae. After departing at 1115 they arrived over Port Moresby an hour later. More than two dozen Airacobras were scrambled and ten of these made contact with the Zeros, with a large-scale combat evolving some 25 miles to the west of Port Moresby. The engagement was keenly fought, with the Zeros expending over 4,000 rounds of 7.7mm and 20mm ordnance. The Tainan *Ku* claimed ten kills, with four

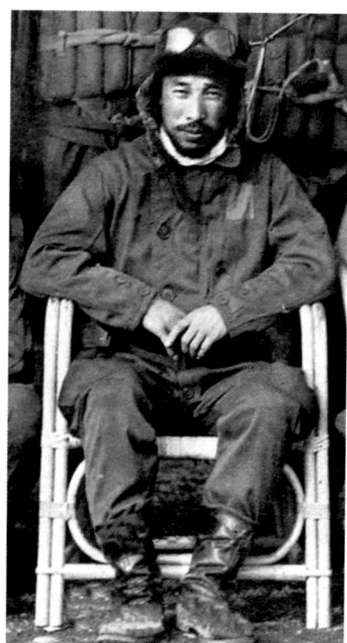

Tainan Ku hikotaicho Lieutenant-Commander Nakajima Tadashi at Rabaul in mid-1942. Nakajima led 21 Zeros on a fighter sweep over Port Moresby on 25 June.

Catalina A24-29 undergoes outdoor maintenance at Bowen. The small force of RAAF Catalinas flew many long-range night bombing missions against Japanese targets during June-September 1942.

Catalina PBY-5 A24-21, No. 11 Squadron, RAAF, Bowen, Queensland, June 1942.

of them awarded to FPO1c Nishizawa Hiroyoshi, who flew on Nakajima's wing. However, the only loss to either side was a 39th FS P-39F flown by Lieutenant Robert Rose which was written off when landing with combat damage at Twelve-Mile.

The following day, 26 June, saw a significant daylight raid on Port Moresby. This comprised 20 Genzan *Ku* G3M Nells led by *buntaicho* Lieutenant Ishihara Kaoru. The bombers joined an escort of eleven Lae-based Tainan *Ku* Zeros over Buna before proceeding to Port Moresby. Ishihara found thick cloud over the town with only a few patches of clear sky to assist him in identifying the target area. Ishihara's Nells carried a mixed bomb load of 69 x 60-kilogram bombs and twenty heavier 250-kilogram bombs. Not surprisingly in these conditions the bombs fell into foothills three miles north of Seven-Mile and did no damage.

Some 26 Airacobras were scrambled and these were bounced by the Zeros under the leadership of the experienced and aggressive Lieutenant Kawai Shiro. In this combat the Japanese claimed three P-39s while the Americans claimed one Zero, but just one aircraft was lost. This was a 40th FS P-39 flown by Lieutenant William Stauter which went missing in action.

A short time later five of the Airacobras caught up with the Nells as they passed over the mountains near Kokoda and made several attacking passes. The Nell gunners fired 2,441 x 7.7mm rounds in self-defence. This was a brief but intense fight. The Nells later submitted they had been attacked by five fighters, claiming one shot down and one probable. Eight Nells were hit by American bullets. One Nell sustained direct hits to the cockpit and made a forced landing at Lae with all eight crewmen wounded, three seriously including pilot FPO1c Hirayama Jinro and his co-pilot. The Nell was written off. Another damaged Nell followed Hirayama's bomber into Lae fifteen minutes later.

During the day three RAAF Catalinas, led by Squadron Leader Francis Chapman, were each loaded with ten 250-pound bombs at Noumea, where they had arrived the previous day from Bowen. That night two of the flying boats raided Tulagi, some 1,000 miles distant, after refuelling at Havannah Harbour, Efate, in the New Hebrides. The first Catalina, A24-14 piloted by Chapman, returned direct to Noumea, but the other, A24-17 flown by Flying Officer Robert Seymour, decided to return to Havannah Harbour to refuel.

As Seymour made his landing approach, a USMC Wildcat from VMF 212 mistook the red inner roundels on the Catalina's wing for Japanese insignia and opened fire, causing damage to one of the wings before Seymour was able to make a safe landing. Seymour reported:

> … with the harbour in sight and on descent about ten miles out, we were attacked by an American Wildcat … it had not been informed of our presence … we were subjected to three diving attacks from vertically overhead. Most of the 0.50-calibre rounds passed us, but there were some hits on each pass. Both tanks were holed, the port aileron was shot away, the hull was holed in seven or eight places, and the wings in about twenty.

The damage was later repaired at Noumea by the seaplane tender USS *Curtiss*. Subsequent to this incident the RAAF painted out the red roundels so that its aircraft were identified by a blue and white insignia.

Early the following morning, 27 June, the third Catalina at Noumea, A24-12 flown by Flight Lieutenant William Bolitho, made its own attack on Tulagi. Three nights later Squadron Leader Chapman made another such raid. After again refuelling at Havannah Harbour he returned direct to Bowen from Tulagi.

Meanwhile a fatal accident had occurred on 26 June, when Wirraway A20-84 of No. 24 Squadron crashed into Mount Louisa, not far from RAAF Townsville, with the loss of both crewmen.

A pioneering night operation was finally flown in the evening of 27 June by five B-26s from the 2nd and 408th Bombardment Squadrons, however the mission has a curious background including a first failed attempt at the concept. This required instrument flying, a skill at which

An F1M Pete floatplane taxies towards its anchorage in the Shortlands Islands. Floatplanes from the Kiyokawa Maru often operated from forward locations, including from Salamaua where a Pete searched for the Australian troops responsible for the commando raid on 29 June.

the B-26 pilots were not particularly adept. Furthermore, the B-26 was considered a "slippery ship" and whilst fast, was not a particularly stable platform in instrument conditions compared to a Catalina, Fortress or even a B-25 which was slower and had anhedral wings for stability. Given these circumstances it is unclear why such a mission was authorised for B-26s over the Owen Stanley ranges at night.

The Townsville-based B-26 crews viewed a documentary film on survival tactics for downed aircrews in New Guinea on 23 June, then spent the rest of the night on the town. They flew to Port Moresby during the afternoon of 26 June and had a late dinner at 2300 before departing for their first night mission attempt in the early hours of 27 June at 0030. However, they could not get through heavy cloud over the mountains and returned to Port Moresby. The dead-tired crews waited around Seven-Mile the next day, before finally departing again around 2100. Lieutenant Merrill Dewan was the navigator aboard *Blue Grass Bettye* and wrote:

> This time we got through the pass in the mountains, but the weather over Lae was terrible. In fact we were over the target before we knew it, and AA opened up on us. This marked the field for us so we let 'em have it – 3,000 pounds of bombs. Then all hell broke loose. Our left engine started popping and we commenced going around in a wild uncontrollable circle, towards the ground. [Lieutenant Robert] McCutcheon and [Lieutenant Lomas] May were fighting hard at the controls … all the time we were losing altitude … scared to death but I helped them and we fought the controls like demons for long sweating seconds. Finally at 1,000 feet of altitude we managed to pull her out. We got back across the mountains safely but aged ten years in doing it and landed safely at Moresby with a shot-up engine and an unnerved crew of boys. We were shaking like leaves when we got out of the aircraft…

Dewan's description is a classic case of loss of control in instrument conditions and could have been resolved more quickly in visual conditions. In the face of bad weather, it is a credit to the perseverance of the crews that three B-26s dropped their bombs in the vicinity of Lae while the other two proceeded to the alternate target of Salamaua. While the results could not be observed due to poor visibility, the crews were proud of carrying out the first night mission by

B-26s in the theatre. However, level heads prevailed and this was the last major night mission which Marauders would fly in the theatre.

The 27 June B-26 night raid was the first of several Allied attacks planned to coincide with the imminent land raids on Salamaua and Lae. Later that night a single Catalina spent over four hours above Salamaua while another did the same over Lae making "nuisance raids". At intervals each flying boat unloaded its main armament of 8 x 500-pound bombs, while the crew hand-dropped 20-pound fragmentation bombs and empty beer bottles, which made a loud whistling sound.

The following night, in the early hours of 29 June, Australian troops made their commando raid on Salamaua. During the day, as they retreated into the safety of the jungle a F1M Pete floatplane of the *Kiyokawa Maru* air group made several sweeps over the area trying to identify the evacuation routes taken by the fleeing raiders. This aircraft also dropped bombs into Komiatum and Mubo villages which were on the main mountain trail used by the Australians. Also, a Mavis flying boat was seen to land offshore Salamaua and discharge 40 troops, the first reinforcements of many which would arrive there in coming days.

On the evening of 29 June some nine B-17s ventured out for another night mission to Rabaul. At least two of the Fortresses found the target area and dropped bombs on the wharf area, where one fire was started, and on Lakunai strip. Searchlights and AA fire were encountered. Two other B-17s attacked Lae, presumably being the alternate target.

Meanwhile the IJA had requested information on a possible land route over the Owen Stanley mountains to Port Moresby, and on 28 June the first of several reconnaissance missions was flown over the area by a Tainan *Ku* C5M Babs, with a Zero escort. Then, on morning of 30 June another Lae-based Babs escorted by four Zeros led by Lieutenant Kawai Shiro returned to the same area on a two-and-a-half-hour mission and took more photos. The encouraging results resulted in a repeat return to the area by a solitary Babs that same afternoon. These flights resulted in detailed reports being submitted to IJA headquarters in Rabaul.

*

As noted in the overview, one of the four squadrons of the 19th BG, the 435th BS, was largely tasked with reconnaissance. This unit had not fought in the Philippines and Java with the other three squadrons of the group but had arrived in Australia via Hawaii in February as an independent squadron, before being incorporated into the 19th BG. Over time it came to specialise in the long-range reconnaissance role, using photo-equipped B-17Es, and became popularly known as *The Kangaroo Squadron*. Squadron members accordingly designed their own unit patch featuring a hopping kangaroo (signifying Australia) with a looking glass at its eyes (the reconnaissance role) and its tail holding a bomb (secondary bombardment role). In the background was a cloud, which was often the best defence of solo reconnaissance aircraft.

By mid-1942 the 435th BS was serving a function of considerable strategic importance. With the USN planning to land at Tulagi (and subsequently Guadalcanal) in early August, the 435th BS was the primary means of obtaining reconnaissance of the Solomons and surrounding waters (there were not yet any long range land-based aircraft operating from the New Hebrides, where

Liberator LB-30 serial AL515 conducting reconnaissance over Tulagi in July 1942. Inset: Insignia of the 435th Bombardment Squadron, which became popularly known as The Kangaroo Squadron.

airfield construction was still underway). To improve the long-range reconnaissance capability, in June the squadron was allocated three Liberators. These were early model LB-30s, originally built for Britain but impressed for USAAF service just after Pearl Harbor.

The three LB-30s in question had British serials and were the survivors of an original twelve 7th BG Liberators that had been flown to Java in the first weeks of 1942. These bombers bore the most curious of markings for the Pacific theatre. They had been accepted into the USAAF inventory still bearing the British RAF night Bomber Command paint scheme, overall matt black with camouflaged upper surfaces. The first of these, AL515, reached the 435th BS at Townsville on 19 June, the second (AL573) arrived on 28 June and the third (AL570) arrived on 13 July. The Liberators had a much longer range than the B-17Es and using auxiliary fuel tanks could fly for up to twelve hours. Within days of its arrival, AL515 had been used to overfly Truk, the first such reconnaissance of the major IJN base since RAAF Hudsons had flown there in January from Rabaul.

However, a drawback was that the LB-30s didn't have superchargers and thus had a modest operational ceiling of only some 12,000-feet. Hence the B-17Es were preferred to overfly defended locations at high altitudes of up to 30,000-feet or more, but the LB-30s would soon be busy over locations such as the lower Solomons where there were no land-based fighters.

Fortresses of the 435th BS flew reconnaissance sorties most days. During the last week of June common targets included Rabaul, Lae, Kavieng and Buka, with airfield construction activity noted at the last two locations. On 13 June a B-17E overflew Guadalcanal at 20,000 feet but in conditions of "dense overcast cloud and rain" nothing was seen. The furthest location reached by the Fortresses appears to have been Kapingamarangi, located about half-way between Rabaul and Truk, where newly built jetties were noted on 21 June. The first mission by a LB-30 was the flight to Truk on 25 June, but no sightings of interest were reported.

A 22ⁿᵈ BG Marauder makes a low-level pass against Lae, approaching from the east over Huon Gulf, and is intercepted by Tainan Ku Zeros.

CHAPTER 3

INDEPENDENCE DAY
REGIONAL OPERATIONS 1–7 JULY

The raid by Australian troops against Heath's Plantation near Lae unfolded in the early hours of 1 July. From around 0500 that morning some coordinated air support was planned to cover the withdrawal of the Australians. Two RAAF Catalinas dropped a mix of 20- and 500-pound pounds on Lae's airfield, while formations of B-25s and B-26s bombed the town and a ship at the wharf. However bad weather hampered the attacks, with three out of six B-26s and a third Catalina unable to find the target. During the day six more B-26s bombed Salamaua, while two B-17s couldn't find Lae due to heavy overcast: one jettisoned its bombs 50 miles from the target area and the other returned with its bombs to Port Moresby.

That morning three Zeros strafed the road out of Lae but none of the Australians were hit. The Australian raids resulted in a sharp Japanese response, both in the air and on the ground. Among the ground reinforcements available to aggressively hunt the Australians in coming days was a detachment of the 5th Sasebo SNLF which had arrived at Lae from Truk on the *Kinai Maru* on 30 June. This was the ship targeted by the B-25s early that morning. It was undamaged and sailed for Truk on 1 July, arriving two days later.

However, it was obvious to the Japanese that the Australian raiders had come from the mountain strongholds of Wau and Bulolo. On 2 and 3 July a *chutai* of nine No. 4 *Ku* Bettys, each carrying 12 x 60-kilogram bombs, targeted these centres. At Bulolo a building used by the army as a store house received a direct hit and two soldiers were killed. At Wau six houses were destroyed and a number of natives were wounded. The incident adversely affected the morale of natives which the Australians depended on to carry supplies and many, terrified by the bombs, deserted.

After these raids some Bettys returned to Vunakanau while others landed at Lae and then made further raids before returning home. Among the targets was Gabmatzung Mission which was further upstream along the Markham River from Heath's. These raids were accompanied by strafing attacks by Zeros. Although there were some close escapes, there were no Australian casualties as a result of these attacks.

Despite the Allied air attacks on Lae, the Tainan *Ku* detachment there remained as busy as ever. On 2 July pilots flying defensive Combat Air Patrols flew 28 sorties and another 26 sorties the following day. Two more C5M reconnaissances of the Kokoda area were also flown on 3 July with escorting Zeros. During one of these escort missions, after the C5M turned back for Lae, *shotaicho* Lieutenant Kurihara Katsumi could not resist leading his *shotai* of three Zeros on a sweep over Port Moresby. A brief attacking pass on three Airacobras was made, but the Americans were able to take evasive action. Kurihara then led his trio back to Lae.

On 3 July Port Moresby was a hive of activity as bombers flew up from Queensland for a maximum effort against Lae planned for 4 July, American Independence Day. These included sixteen B-26s in what would be one of the largest daily combinations of Allied bombers to take to the skies in weeks. On the evening of 3 July, a 22nd BG Marauder pilot wrote in his diary that:

It's going to be a smashing attack ... some party!

Earlier that day three B-17s had flown another bombing mission against Lae, with each carrying two immense 2,000-pound bombs. Only two of the Fortresses reached the target area and recorded their bombs hitting near the runway. However, the huge explosions didn't prevent eight Zeros from scrambling. These caught Lieutenant Clyde Kelsay's Fortress about 90 miles south of Lae, and during the course of several attacking passes they peppered the bomber with fifteen bullet holes. Two of the B-17 gunners were wounded.

The first of the 4 July attacks comprised seven 3rd BG B-25s which struggled to find Lae in murky conditions between 0400 and 0500. Eventually two pilots were able to drop their bombs over the runway and dispersal areas in the face of heavy AA fire. However, two B-25s were unable to find Lae and jettisoned their bombs, while another bombed the secondary target of Salamaua. As the B-25s sped home they weren't pursued by Zeros but were met by Airacobras at a planned rendezvous near Cape Ward Hunt.

The RAAF also contributed to these raids. Four Catalinas departed Cairns during the afternoon of 3 July and were above Lae 'drome individually during a two-hour period early the following morning. The flying boats targeted the runway and dispersal area and dropped a mix of 20- and 250-pound bombs, with some of the latter fused with six- and twelve-hour delays.

Meanwhile, from Horn Island four No. 32 Squadron Hudsons set out to attack Salamaua in a change from their normal reconnaissance operations. Taking off during the evening of 3 July, they were led by the squadron CO, Wing Commander Derryck Kingwell, who reported one bomb landed accurately on Salamaua's small airstrip. The Hudsons also strafed buildings. However, one of the Hudsons, A16-193, piloted by Flight Lieutenant Pat McDonnell, was lost when it never returned from the raid.

The B-26 attack was planned to be in two eight-aircraft formations, hitting Lae towards mid-morning and then two hours later. The idea was that the second wave might catch Zeros on the ground refuelling after scrambling and pursuing the first wave. However, shortly after dawn bad weather reports delayed the planned departure and Japanese actions then interfered with the mission. The first interference had been a night raid at 0357 which put Seven-Mile on full alert. This was a pair of No. 4 *Ku* Bettys which dropped patterns of 60-kilogram fragmentation bombs accurately, landing just 225 yards from the main runway but doing no damage. In fact, six Bettys had set out from Vunakanau but only one pair was able to find Port Moresby in the dark and cloudy conditions.

The second interference was more substantial, with the detection of twenty incoming Zeros at 1015 again putting Seven-Mile on full alert. The Zeros benefited from the leadership of three experienced *buntaicho*. Leading the full strength (nine-strong) No. 1 *Chutai* was Lieutenant

The mountain township of Bulolo just before the outbreak of war. It was bombed by No. 4 Ku Bettys in early July as retribution for the Australian raids on Salamaua and Heath's Plantation.

Kawai Shiro, while Warrant Officer Takatsuka Tora'ichi led the six strong No. 2 *Chutai* and Lieutenant Sasai Jun'ichi led the five strong No. 3 *Chutai*. Sasai's *chutai* remained in reserve over Kokoda during the forthcoming engagement, the idea being they could intercept any bombers heading for Lae. In the event they saw no action. Takatsuka, a 28-year-old China veteran, was a new Tainan *Ku chutaicho* who had only joined the unit June. His age and experience had quickly boosted him into the ranks of the "most-trusted" among both the Tainan *Ku*'s other pilots and its leadership.

With the first eight Marauders fully fuelled and armed, confusion reigned as they made emergency take-offs to orbit some miles out to sea over the Gulf of Papua. Standard procedure for was the bombers to wait until the raid was over before returning to Seven-Mile. At the same time around a dozen Airacobras were scrambled but most of these failed to make contact with the enemy.

A patrol of thirteen 39th FS Airacobras, led by the unit CO Major Jack Berry, was already airborne at 21,000 feet. These were bounced by the Zeros which attacked from "great altitude" in the vicinity of the 'drome at Rorona (30-Mile). Likely shot down in these first passes were Lieutenants James Foster and Frank Angier, both of whom bailed out safely.

After the remaining Airacobras dived away into cloud to evade the Zeros various low-level dogfights developed over a wide area. One other Airacobra was lost, after Lieutenant William Marlott made a forced landing near the coastal village of Boera, with two others sustaining damage.

All twenty Zeros had returned safely to Lae by 1215, with the exception of one which diverted via Salamaua. The Tainan *Ku* pilots claimed a dozen P-39s destroyed, which was a four-fold over-claim. However, they did not have to wait long for further action.

While the Zeros were still busy over Moresby, Marauder pilot Major Brian "Shanty" O'Neill decided to head directly to Lae before the fighters could return there. He was joined by two others which formed up with him. On approach to Lae at 13,000 feet, they descended to 8,000 feet before dropping their loads of 100-pound bombs and making a speedy escape. They were immediately harassed by a *shotai* of three Zeros which followed them for twenty minutes. O'Neill decided to bypass Port Moresby altogether on the return flight and all three B-26s made it to the safety of Cairns and Townsville.

Back at Seven-Mile, the second group of eight B-26s departed for their mission just as the Zeros had departed. This was a 22nd BG headquarters detachment of four aircraft led by Lieutenant Walter Krell and a 33rd BS quartet led by Lieutenant George Kahle. As both groups approached Lae Krell was annoyed that Kahle had not followed him closely but was instead trailing a mile behind. After bombing from 7,000 feet Krell made a wide arc so that Kahle might catch his formation and provide for mutual defence. Krell noted that:

> Kahle couldn't have set himself up for a worse situation. By lagging through the target, he gave the Zeros every chance to get set, and they were on him like hornets.

Nevertheless, the eight B-26s managed to combine together and dive away at high speed as the Zeros made their attacks. The B-26 gunners were busy as the Marauders hit a speed of 340 miles per hour before levelling out a sea level. During this descent Krell recalled:

> One Zero attacker seemed unusually persistent and as I turned into him to spoil his interception, he kept pulling down into us. So, as he passed over my left wing, he was nearly inverted with all guns blazing. An instant later, this Zero had rammed Moe Johnson's B-26 in Kahle's Number Four position.

Both the B-26 and the Zero crashed into the ocean with the loss of all onboard both machines. The Zero pilot was Flyer 1c Suizu Mitsuo, and among his No. 3 *Chutai* comrades was the famed ace FPO1c Sakai Saburo, who later wrote in his memoir that he had a premonition the collision involved Suizu:

> Back at base we assembled in front of the command post. More than anything we wanted to know who the [missing] pilot was. I thought it would be Suizu. At Lae he was one of the least experienced pilots. Every day he was one of those who witnessed the more experienced ones departing for combat. In his view he too was a fighter pilot, but often they would not let him go. He was unhappy. And he used to say "One day I will do something extraordinary. Next time I will shoot down a bomber even if I have to collide with it."

In total the B-26s had loosed 272 x 100-pound bombs over several targets, including the dispersal area, AA positions and a suspected headquarters building. A fire and thick black smoke were reported to be coming from a hangar. Japanese records suggest the B-26s likely ran

Martin B-26 Marauder, serial 40-1411, MARTINSMISCARRIAGE, 19th BS, 22nd BG, Woodstock, Queensland, July 1942.

The day after the Independence Day raids, several B-26s attacked Salamaua. After returning to North Queensland that afternoon this Marauder, 40-1533, ran off Cairns' runway after suffering an engine failure and was written off.

into dribs and drabs of Zeros returning from Port Moresby which were low on both fuel and ammunition. Aside from Suizu's effort, there was little evidence of determined attacks against the bombers on this occasion.

Late that afternoon a force of six B-17s which had departed Horn Island also attacked Lae 'drome

Australian soldiers survey the damage to an Airacobra at Port Moresby after another attack by No. 4 Ku Bettys. An Airacobra and two B-26s were destroyed as a result of the 5 July raid.

from 28,000 feet. The Fortresses dropped 84 x 300-pound bombs which landed across the runway and dispersal areas. Riding in one of the bombers was war correspondent Howell Walker from the *National Geographic* magazine who observed the bombs setting fire to a fuel dump:

> A tremendous fire blazed on the edge of the aerodrome and masses of black smoke billowed up from two other fires in the same area.

Several minutes later the bombers were intercepted by five Zeros which made frontal and stern attacks which resulted in light damage to one Fortress. Walker later described the Zero attack:

> Their tracer shots poured past our windows like a horizontal hail of red-hot rivets. Never before had I been in actual combat. Now that I was, it did not seem real.

On landing at Horn Island the undercarriage of one B-17 collapsed on landing and punched upwards through the port inner engine, just behind the firewall. The damage grounded the bomber for some time.

During the first week of July the handful of Tainan *Ku* C5Ms were also busy. On 3 July a Babs escorted by three Zeros found a bridge near Kokoda. This was Wairopi on the Kumusi River, so named because of the "wire rope" bridge, and the report provided further encouragement to the IJA that a road of sorts ran between the coast and Kokoda. Fortresses had used Horn Island as an advanced base several times in the last few days and probably for that reason a C5M overflew the island on 6 July. This was reported from Horn Island as a single "Zero" overflying the base at 10,000 feet at 1043 on 6 July.

The Japanese were quick to respond to the 4 July raids, hitting Port Moresby on both 5 and 6 July. The first raid saw twenty No. 4 *Ku* Bettys departing Vunakanau at 0645 on 5 July led by *hikotaicho* Lieutenant Commander Watanabe Hatsuhiko. After making rendezvous with fifteen Lae-based Zeros over the north Papuan coast, the formation entered Port Moresby airspace at 0955. The Nos. 2 and 3 *Chutai* broke away and made their own attack at 1010 from 23,000 feet. The pattern of bombs landed among dispersal bays at the south-east end of the Seven-Mile runway. One soldier was killed and three others were injured.

Fifteen minutes later the No. 1 *Chutai* aircraft made another run over Seven-Mile, dropping their bombs accurately among dispersal bays on the south-western side of the runway where a number of B-26s were parked. Two of the bombers were destroyed, one because of heat damage and another due to shrapnel. A single P-39 was also written off and a small fuel dump went up in flames. In total the Bettys had unloaded 19 x 250-kilogram and 106 x 60-kilogram bombs. The Bettys were engaged by AA guns and two received slight damage.

While this raid was underway two flights of B-26s, each of four aircraft, were in the air. Unable to reach Lae due to inclement weather they bombed Salamaua instead. Later one of these B-26s (40-1533) was written off in a landing accident at Cairns following an engine failure. Earlier on this day several aircraft from Horn Island had made night attacks: two trios of B-17s from Horn Island had bombed Lae's runway while a single Hudson had bombed Salamaua.

No. 4 *Ku* repeated the effort against Port Moresby on 6 July. Underlining the importance of the mission was the fact that *hikotaicho* Lieutenant Commander Watanabe Hatsuhiko was once again personally leading the mission – for the second time in two days. This time the force was 21 Bettys which unloaded a mix of over one hundred 100- and 250-kilogram bombs over

Mitsubishi F1M2 Pete CN 157, RI-19, Kiyokawa Maru, July 1942.

Seven-Mile at 0925. The bombing was less accurate than on the previous day, hitting a ridge alongside dispersal bays. There were no casualties, but a dump of 36 x 100-pound bombs was destroyed. AA batteries were active and fired 221 rounds at the intruders.

The bomber commanded by FPO1c Shinomiya Sousaku experienced engine trouble and lagged the main formation, such that he nonetheless bravely made a solo bombing run almost half an hour later. Some seventeen Airacobras had already scrambled. Eight 39th FS machines remained above Port Moresby while nine from the 40th FS chased the main formation northwards over the mountains. These Airacobras found the Bettys at 21,000 feet over Kokoda while the fifteen escorting Zeros were initially caught unawares. The Bettys reported being attacked by four P-39s and responded with 7.7mm and 20mm defensive gunfire. After landing back at Vunakanau, the Japanese tallied damage to fifteen of the bombers, although some of this was probably caused by AA shells over Seven-Mile. None of the bombers were lost but four crewmen were slightly injured and another was killed as a result of the combat.

However, it was now the turn of a quartet of Zeros commanded by *buntaicho* Lieutenant Sasai Jun'ichi to pursue the Airacobras. The American pilots responded by diving away at maximum throttle and reached speeds of 350 miles per hour. A version of this engagement is recorded in Sakai Saburo's memoirs, whereby after the P-39s scatter in all directions he finds himself chasing one over the ocean north of Papua. Initially the Airacobra is fleeing in the wrong direction, and when the pilot is forced to turn back it gives the Japanese an opportunity to open fire. As the P-39 races back over land the pilot suddenly bails out at very low altitude. With the parachute failing to fully open, the pilot, was killed on impact. This was Lieutenant Howard Welker whose body was recovered and buried at the Gona Anglican mission.

On 7 July, for the third day in a row, No. 4 *Ku hikotaicho* Lieutenant-Commander Watanabe Hatsuhiko was once again leading his men into the air. This time the target was distant Horn Island, with the round trip taking a lengthy nine hours. At 0708 that morning 18 Bettys departed Vunakanau, but three hours later two bombers had been forced to abort included Watanabe's.

Meanwhile, the mission was preceded by a Rabaul-based C5M Babs which had refuelled at Lae before dawn at 0508. The pilot was FPO2c Shimizu Eisaku and the commander was Lieutenant Kitsuka Shigenaga, riding in the observers' seat. Kitsuka also was the deputy commander of the Tainan *Ku* reconnaissance detachment, which consisted of about twenty airmen.

At 1025 the Babs confirmed to the incoming Bettys that a B-17 and two B-26s were present at Horn Island. The "B-26s" were more likely Hudsons of No. 32 Squadron. Meanwhile the approach of the Betty formation was tracked on the ground at Horn Island which enabled the aircraft to be flown away. At 1155 the Bettys bombed the base with what were reported as "180-200 anti-personnel bombs". Little damage was done but four personnel were slightly wounded.

Operations in the Solomons rounded out air activity for the first week of July, where a new Japanese player had just entered the theatre. Already established in the theatre with its H6K Mavises, the Yokohama *Ku* had meanwhile established a fighter detachment in April 1942.

An A6M2-N Rufe floatplane moored at Tulagi, assigned to the buntaicho of the Yokohama Ku floatplane detachment, Lieutenant Sato Ri'ichiro.

It first operated briefly around Rabaul in June 1942 with sixteen A6N2-N Rufe floatplanes. Its *buntaicho* was Lieutenant Sato Ri'ichiro, reassigned from the Yokosuka *Ku* to become CO of the detachment on 15 May. The unit had experienced pilots and was now redeployed to Tulagi with orders to cover the IJN convoy which would offload the supplies required to build an airfield on Guadalcanal. The detachment's 36-strong groundcrew contingent was shipped ahead of the floatplanes to Tulagi on 1 July aboard the *Mogamigawa Maru*, escorted by the destroyer *Akikaze*.

On 3 July Flying Officer Robert Seymour's Catalina A24-15 of No. 11 Squadron, RAAF, bombed Tulagi having departed Noumea and refueled at Havannah Harbour. He then flew directly to Bowen. The next evening Flight Lieutenant William Bolitho in Catalina A24-12 repeated the mission. Meanwhile by 5 July a dozen of the Yokohama *Ku*'s Rufes had arrived at Tulagi from Rabaul from where they conducted their first sortie: a pair of Rufes led by FPO1c Hori Tatsuo flew a defensive patrol from 0710 to midday.

Then at 1620 the following afternoon a LB-30 Liberator on a reconnaissance mission from Townsville reported a cruiser, two destroyers, four transports and various small craft off Guadalcanal. The Liberator dropped two 500-pound bombs from 6,500-feet, and noted two airborne floatplanes, being Tulagi-based A6M2-N Rufes. The Liberator in a sense was lucky. Although it spotted only two floatplanes there were in fact a dozen, split into two echelons, just returning from an extended shipping patrol which had departed Tulagi at 0600 that morning. *Buntaicho* Sato and five accomplices climbed to try and reach the bomber while the other detachment of six, guarding scattered groups of ships, missed the action. Sato logged the "unsuccessful" encounter at 1630 local time, and all his floatplanes had returned safely to Tulagi by 1750. Sato submitted that the "large aircraft" got away only as its speed was approximately 20 kilometres per hour superior to his floatplanes. It appears the two floatplanes spotted by the LB-30 were Sato and his unidentified wingman who led the brief attack.

The vessels comprised the IJN convoy carrying out Operation SN, which was the establishment of the much-invested airfield on Guadalcanal. The ships included light cruiser *Yubari*, four destroyers, ammunition ship *Kotoku Maru* and IJN transports *Azuma, Azumasan, Hokuroku, Kanto, Meiyo* and *Matsumoto Marus*. The ships were carrying over 2,500 men of the 11th and 13th Establishment units, SNLF marines, airfield construction supplies and drums of aviation fuel. Some of these vessels had originally been loaded with these supplies intended for Midway Island but had now been redirected to the Solomons after the failure of the Midway operation. The convoy had departed Truk on 29 June and began unloading at Guadalcanal on 5 July. This was done with only two transports unloading at a time to minimise the risks of Allied air attack. Airfield construction work would not begin until 16 July.

At 1055 the following morning, 7 July, a second 435th BS LB-30 Liberator from Port Moresby was tasked to locate and attack enemy transports in vicinity of Tulagi, and reconnoiter various nearby locations including Guadalcanal. It found the same two large cargo vessels north of Guadalcanal seen the previous day which it attacked at 1055, claiming a direct hit on small motor torpedo boat. Its crew also noted "two single float low wing aircraft - made no interception." This pair were again drawn from *buntaicho* Lieutenant Sato Ri'ichiro and his floatplanes, also busy this day. With one floatplane ashore with a mechanical problem from the day before, on 7 July Sato was tasked with ongoing coverage of the convoy unloading off Guadalcanal. He tasked Lieutenant Ebisuda Yoneo to conduct the first shift with three Rufes. These departed Tulagi at 0650, then loitered over the convoy for nearly ten hours with throttles set to maximum endurance, before returning safely. Sato led a second patrol of eight which departed at 0710. They returned to Tulagi after an eight-hour patrol over the ships.

Meanwhile, light AA fire was experienced after the Liberator again dropped two bombs. These two Liberator missions had confirmed Japanese intentions to construct an airfield on Guadalcanal, and within days USN plans to invade Tulagi in August were modified to include Guadalcanal as well.

Buntaicho of the Yokohama Ku fighter detachment, Lieutenant Sato Ri'ichiro, who doggedly led the defence of Tulagi and Guadalcanal with his force of Rufe floatplane fighters during July and early August.

An RAAF Hudson turns for its final approach as it brings in supplies to Kokoda.

CHAPTER 4

THE LULL BEFORE THE STORM
REGIONAL OPERATIONS 8–20 JULY

The thirteen-day period ending 20 July saw a relative lull in operations by both sides. This was partly due to the need to take a breath after the operational intensity of the first week of the month. For example, no B-26 missions were flown during this period. However, it was also due to unusually bad weather which descended over the New Guinea region at this time. In addition, both sides had an eye on upcoming offensive operations and the need to conserve strength to have maximum airpower available when needed. Other tasks were related to the airfield construction activities underway by both sides. For example, the RAAF Catalina force flew no attack missions during this period but instead was heavily engaged flying anti-submarine protection missions for ships taking troops and materials to Milne Bay.

On 7 July another supply ship had avoided Allied reconnaissance and arrived safely at Lae. This was the sizeable 7,190-ton IJN auxiliary transport *Nojima Maru* which had sailed directly from Truk. Escorted by the minelayer *Tsugaru* and the submarine chaser *CH-31*, the transport docked at Lae at 1730 and unloaded during the night. *Tsugaru* departed for Truk at 2100 that evening, while *Nojima Maru* and *CH-31* sailed for Rabaul on 8 July where they arrived safely the following day.

Less fortunate was the 2,775-ton IJN auxiliary transport *Tenzan Maru* which on 8 July was nearing Rabaul towards the end of a voyage from Japan. When just 30 miles north-west of the port, and despite the presence of submarine chaser *CH-30*, the ship was torpedoed and sunk by the USS *S-37* with the loss of 81 passengers and one member of the crew. Despite a depth charge attack by *CH-30*, the *S-37* was undamaged and managed to get clear. The submarine remained in the area for several more days before technical problems forced its returned to Brisbane where it arrived on 21 July.

Meanwhile some preliminary IJN efforts had been expended to build an airfield at Kieta on Bougainville. On 8 July a Mavis flying boat brought a party of 25th Air Flotilla officers to assess the site. That evening the 25th Air Flotilla commander, Rear-Admiral Yamada Sadayoshi, sent a report to his superiors:

> As a result of this intelligence, it is anticipated that one runway can be completed by early September using the main strength of 14th Establishment Unit at Rabaul. However, even an upgraded airfield could not be used as an emergency strip for medium land attack aircraft owing to the obstructions on the periphery of the runway and the poor drainage. There is no room for expansion of the airbase, so there is absolutely no prospect for using this site according to requirements … additionally, no suitable sites for an airbase have been located on Bougainville.

Following this report the work at Kieta was suspended. It would be some months before any sizeable airfields that could properly handle bombers were constructed in Bougainville or the Solomons (excepting Guadalcanal which would be invaded prior to any Japanese aircraft arriving there). Some development had also been noted at the former Australian airstrip on Buka, but for the present that was only suitable for emergency use by fighters.

The 10 July saw the resumption of attacks against Port Moresby. At 0700 *hikotaicho* Lieutenant-Commander Tsusaki Naonobu led 21 G4M Bettys of the No. 4 *Ku* into the air from Vunakanau, escorted by twelve Tainan *Ku* Zeros from Rabaul. An hour later eight Lae-based Zeros were airborne, which would orbit over Cape Ward Hunt on the northern Papuan coast, where they would act as additional cover should the Bettys be pursued northwards by Allied fighters. This was in fact an identical tactic that had been used successfully by Airacobras to protect B-25s and B-26s after they had bombed Lae. To Allied and Japanese pilots alike, Cape Ward Hunt was an easily recognisable landmark, with a 150-foot high bare rock outcrop and white sandy beach standing out against miles of jungle-fringed coastline.

On this occasion neither the backstop nor the escorting fighters were needed, as the formation had not been detected by the outer arc of spotting stations that normally warned Port Moresby of incoming raids. At 1013 the bombers commenced their attack run against Port Moresby's harbour from 23,000 feet. Eleven 35th FG Airacobras had scrambled, but their poor rate of climb meant they could never catch the intruders with the limited warning provided on this occasion.

The bombers were engaged by two batteries of Australian 3.7-inch AA guns. Despite only having a narrow window in which to fire 48 rounds, the fire was deadly accurate with the first salvos bracketing the lead bomber. The gunners reported the Bettys splitting into three smaller formations as a result and climbing to 26,000 feet. Several of the bombers were then seen to lose altitude and trail smoke.

The Bettys were targeting two ships in the harbour, the corvette HMAS *Bendigo* and the old coal-burning Burns Philp cargo ship *Mangola*. Two patterns of bombs fell about 50-100 yards wide of each ship, which were probably saved by the accuracy of the AA fire which interrupted the bomb run. The standard IJN practice was for the entire formation to drop their bombs on the signal of the lead aircraft, which was that flown by Tsusaki. However, Tsusaki's aircraft had indeed been hit, and the crew was struggling to fight a fire as the pilot bravely held course to target the ships.

The crippled Betty subsequently crashed near the coastal village of Gaile some 35 miles south-east of Port Moresby. Some grisly finds followed, as other Betty crews reported seeing their fellow airmen jumping without parachutes to escape the flames. Several burned and mutilated bodies were found over a wide area, with at least one also showing signs of shrapnel wounds. Some valuable intelligence material in the form of maps and documents was recovered from the wrecked bomber. All twenty of the remaining Bettys returned to base, but seven of them had been damaged by the AA fire to some extent.

Despite the loss of their *hikotaicho*, the battered but determined No. 4 *Ku* returned to Port

Port Moresby harbour during the second half of 1942, crowded with supply ships and several moored RAAF flying boats. On 10 July the No. 4 Ku hikotaicho Lieutenant-Commander Tsusaki Naonobu led an attack against ships in the harbour but was shot down by AA fire.

Mitsubishi G4M1 Model 11 Betty F-313, No. 4 Ku, Vunakanau, July 1942.

Moresby the following day, 11 July. Once again 21 Bettys departed Vunakanau, this time led by Lieutenant-Commander Watanabe Hatsuhiko. The Bettys were followed by a dozen Tainan *Ku* Zeros 30 minutes later, who were due to rendezvous with the bombers over Cape Ward Hunt.

However, plans went awry when at 1015 the Zeros sighted six 19th BG Fortresses on a reverse heading about 100 miles north of Buna. These had departed Port Moresby and were on their way to bomb Rabaul. Lieutenant Sasai Jun'ichi led his half dozen No. 2 *Chutai* Zeros away from the Bettys to pursue the American bombers. The Fortresses jettisoned their bombs and turned for Horn Island, after which a running battle ensued for ten minutes. On this occasion the American defensive tactics were effective, and they shot down FPO3c Suzuki Matsumi. Despite the other Zeros expending most of their ammunition, the Fortresses sustained only minor damage and no casualties.

While Sasai's *shotai* managed to re-join the six Zeros of the No. 1 *Chutai* after battling the Fortresses, the other two Zeros returned to Rabaul. At 1038 the nine Zeros were seen orbiting Cape Ward Hunt by an Australian observation post, but they had lost coordination with the Bettys. The Bettys had either waited at the wrong rendezvous point, or been delayed by weather, as it was over an hour later when they emerged over Port Moresby at 1150 without their escort. By this time just twenty Bettys were in the formation as one bomber had aborted and landed at Lae.

Having received some warning of the bombers' approach, 23 Airacobras had scrambled and tried to climb as fast as possible. However, this was too much for some of the worn engines. A 40th FS P-400 flown by Lieutenant Edward Gignac experienced engine failure on reaching 15,000 feet. He made a forced landing at high speed near the 30-Mile 'drome (Rorona) but was seriously injured when his Airacobra broke up. Seven other American fighters were forced to return to base with engine trouble, while another five never made contact with the enemy.

As the Bettys made their bomb run, they once again targeted the *Mangola* which was still in the harbour after being bombed the previous day, about half unloaded. The Naval Officer in Charge of Port Moresby noted in his report that:

> Nineteen heavy bombers proceeded up harbour from seaward and pattern bombed the *Mangola* at 1100 [local rather than Japanese time]. The vessel was hidden from sight by the bomb explosions, but miraculously was only hit by flying bomb splinters, there being many small holes in the hull and superstructure, but practically no damage and there were no casualties.

However, the ship had a largely native crew, six of whom had jumped overboard with fright during the attack. Subsequently the captain did not feel he could be responsible for their actions and the ship sailed for Townsville the following day, with much of its cargo still onboard.

As on the previous day, the Bettys had been accurately engaged by AA guns which managed to fire 95 rounds as the bombers were over the harbour at 23,600 feet. This time the gunners claimed one definite hit and saw a bomber losing height. They had indeed hit the Betty flown by Warrant Officer Ogawa (first name undecipherable), which was damaged and broke away for Lae with wounded crewmen aboard.

As the Bettys departed the area, five Airacobras finally managed to make attacking passes which did some minor damage and left eight crew slightly wounded. This was a fierce fight, for the Betty crews later reporting being attacked at 8,000 metres by fourteen Airacobras and claimed one shot down. The Betty gunners expended 2,397 x 7.7mm rounds but just 144 x 20mm rounds, suggesting the Americans made beam or frontal attacks to avoid the rearward-facing 20mm cannons. At this point the Zeros belatedly entered the fray, driving off the Airacobras and shepherding the Bettys, six of which had sustained gunfire damage. Seventeen bombers had landed safely at Vunakanau by 1520, with another three making precautionary landings at Lae and Gasmata.

A 41st FS Airacobra undergoes armament maintenance at Seven-Mile. During the second half of 1942 the hard worked Airacobra fleet required increasing workloads to maintain effectiveness of engines and weapons.

The sole loss to either side in the fighter combat was Lieutenant Orville Kirtland of the 40th FS who was missing and has never been found. He was likely a victim of either FPO1c Nishizawa Hiroyoshi or FPO3c Uto Kazushi, both of whom claimed a P-39. Port Moresby would now enjoy a respite of nine days without any raids.

On 11 July two No. 32 Squadron Hudsons departed Port Moresby to bomb and strafe Pora Pora village, on New Britain. This followed an unsuccessful such mission two days earlier when the Hudsons arrived too late in the day to make their attack. Such missions were likely ordered in an attempt to secure the loyalty of resident natives to parties of coastwatchers in the area.

On 13 July a 435th BS B-17E, piloted by Lieutenant Donald Tower, was making a routine reconnaissance flight over Madang, Lorengau, Kavieng and Rabaul. Likely alerted by Madang or Lorengau and aware that the Fortress might overfly their own base, from 0730 seven Zeros led by FPO1c Ota Toshio took off from Lakunai to try and intercept it. Later that morning the Zeros came upon the Fortress over the New Britain coastline due west of Rabaul and made persistent attacks over the course of 40 minutes. While the bomber received hits to its wings, bomb bay, cockpit and outer starboard engine it eventually managed to escape into cloud.

Two airmen pose in front of the wreckage of Lieutenant Curtis Holdridge's B-17E which crashed into mangroves while taking off from Horn Island during a storm on 14 July.

In the early hours of 14 July, six 30th BS Fortresses took off from Horn Island in the dark to bomb Lae. There was rain in the area and intermittent thunderstorms which were impossible to see in the dark. After the first Fortress had taken off around 0325, within minutes the fourth and fifth departing bombers hit a downburst from a thunderstorm cell as it encroached on the airfield.

The first bomber to go down was B-17E 41-2655 flown by Major Paul Lindsey and the second was B-17E 41-2636 flown by Lieutenant Curtis Holdridge. Lindsey got airborne at 0330 with most of the crew secured in the radio compartment for take-off. It ditched violently in the ocean about a mile offshore. The fuselage broke in half from the impact, trapping three crew inside, who were killed. Just behind Lindsey, Holdridge's bomber was also hit by a thunderstorm downburst just as it passed over mangroves at the end of the runway, and it crashed about 300 metres offshore. The starboard wing was ripped off, driving the rest of the airframe into the coral outcrop.

Due to static and noise of the storm, which also jammed radio transmissions, and the fact that two B-17s had crashed within minutes of each other, there was confusion on Horn Island as to what had happened. The other airborne Fortresses proceeded to the target, unaware of the losses. At around 0500 an Australian crash boat located several of the floating crew and rescued them, and also retrieved one of the bodies. The crash boat continued searching until all of the survivors had been rescued.

Hudson Mk IIIA, A16-173, No. 32 Squadron, RAAF, Horn Island, July 1942.

Just a few hours later another aircraft loss occurred in North Queensland. This was Lockheed Lodestar C-56B callsign VHCAD (previously Netherlands East Indies Air Force LT-914) of the 22nd Troop Carrier Squadron. At this stage of the war these Lodestars were making regular runs to Port Moresby, mostly taking ammunition up and bringing wounded back. This particular flight had originated at Essendon (Melbourne) and the aircraft had refuelled at Townsville before attempting the next leg to Cooktown where it was to collect more cargo. Dodging low cloud, the aircraft never arrived in Cooktown and pilot Lieutenant Robert Davis and three passengers were missing. Searches of the coastal area were subsequently made by Wirraways, Catalinas and a B-17 but nothing was found. There is a strong possibility that the wreckage lies in mountains along the rugged North Queensland coastline.

Yet another loss in the area was incurred two days later, on 16 July. This was B-17E 41-2421 from the 435th BS which misjudged a landing approach at Horn Island in darkness that evening, coming in on the wrong side of the flare path. This bomber was flown by Major Clarence "Sandy" McPherson who had departed from Coen bringing five soldiers from the 104th Anti-Aircraft Artillery Battalion for the island's defences. Also aboard was an aircraft salvage crew with heavy replacement parts including landing gear, propeller blades, brake assemblies and specialist tools to repair B-17E 41-2633, which was left damaged on the side of the strip from a separate accident on 4 July. When the bomber made its final approach in the dark, McPherson pulled up sharply to adjust the heading when he realized he was about to touch down on the wrong side of the flare path. The aircraft parts inside were not tied down, and they slid rearwards. With the centre of gravity suddenly moved backwards, the bomber climbed sharply before it stalled. It then fell on its side and dived into the ground before cartwheeling into trees, killing all seventeen aboard.

On 17 July, five B-17Es from the 28th and 93rd BS were over Rabaul mid-morning (one had turned back to Port Moresby with electrical trouble). These bombed the wharf area and recorded one bomb landing on the coaling jetty or a ship alongside it, with others landing near the former Burns Philp wharf and a small vessel nearby. Some 24 vessels of various sizes were seen in the harbour alongside three warships. On this day the Tainan *Ku* flew a total of sixteen Combat Air Patrol sorties over Rabaul. None of the Zeros managed contact with the Fortresses, although the American crews at one stage spotted a single fighter in the air far below them.

On 18 July a trio of No. 4 *Ku* Bettys departed Vunakanau to raid Port Moresby, in company with fifteen Zeros. However bad weather was encountered, and the mission was abandoned. Two days later the No. 4 *Ku* would again take to the skies in one of the strongest raids yet launched against Port Moresby. On this occasion from 0730 that morning 26 Bettys were joined by an escort of fourteen Zeros, with the entire formation soon battling bad weather after leaving Rabaul. At 0900 another six Zeros departed Lae, but it is unclear if they were able to rendezvous with the main formation in the threatening conditions.

By 1025 the bombers had found a gap in the weather and were above Seven-Mile. Some 24 Bettys released their bombs accurately (two others had meanwhile aborted and diverted to Lae) with Hudson A16-179 being destroyed and a dump of 50 fuel drums going up in flames. Bombs also fell close to the 55th Infantry Battalion camp but there were no casualties. Due to the cloud cover only one AA battery was able to briefly open fire and no Airacobras were scrambled.

By 1315 the Bettys had returned safely to Vunakanau, but a drama was just then playing out in respect to eleven of the Zero escorts which had diverted to Lae on the return flight. After refuelling, these took off in small groups for the flight back to Rabaul amid squally weather. Four of these pilots went missing without a trace: Lieutenant Kurihara Katsumi, FPO2c Miya Un'ichi, FPO1c Kobayashi Katsumi and FPO3c Onishi Yoshimi. All four were among a group of ex-Chitose *Ku* pilots which had recently joined the Tainan *Ku* and were relatively unfamiliar with New Guinea conditions. It was the largest single loss due to weather experienced by the IJN in New Guinea to date, although Zero units operating in the Solomons in 1943 would experience similar weather-related losses.

Meanwhile, on this same day two P-39s had attempted a relatively rare reconnaissance flight to Lae, due to cloud cover having rendered recent B-17 flights ineffective. The same weather which had cost the lives of the four Zero pilots as described above now took its toll on one of the Airacobras which became separated in the weather. Lieutenant Frank Angier, who had survived being shot down by Zeros on 4 July, could not find a way home through the cloud and decided to follow the northern coastline. He hoped to find a break in the weather over the Kokoda Gap, but instead the weather worsened, and Angier ran out of fuel. Despite bailing out at very low altitude in the Cape Nelson area, Angier's parachute opened just above ground and he landed uninjured. He was later returned to his unit with the assistance of friendly natives.

Among other operations on 20 July was the operational debut of a new type, the Nakajima J1N1-C Irving. Although originally designed as a long-range escort fighter, the first three to be delivered to the Tainan *Ku* in July were earmarked for reconnaissance operations. The first Irving mission was flown by Warrant Officer Ono Saturo, an experienced fighter pilot, who departed Lae and reconnoitred Horn Island. The flight was correctly observed as an enemy reconnaissance aircraft from the ground at 1026. As will be seen, the Irvings would fly some important missions in the theatre but were sparingly used. This was partially due to the difficulties of maintaining such a small number of this new type in a forward location, but also due to an adequate number of reliable and proven C5M2 Babs already at hand.

During the day the first detachment of 80th FS Airacobras arrived at Port Moresby after being based at Petrie near Brisbane for two months. This was the third squadron of the 8th FG to serve in New Guinea after the other two, the 35th and 36th Fighter Squadrons, had defended Port Moresby in April. Soon the 41st FS would also arrive in New Guinea from Sydney, this being the third squadron of the 35th FG to serve in the theatre. The existing 35th FG squadrons present, the 39th and 40th, would meanwhile fly south to Townsville by the end of July for a rest period.

Also on the move at this time was No. 76 Squadron which had been training on P-40E Kittyhawks at a fighter strip at Aitkenvale Weir in Townsville since April. During June the unit had received its full complement of 24 P-40Es and had 38 pilots on strength. In July the squadron was designated to move to the newly constructed airfield at Milne Bay, and the ground personnel only left Townsville on a ship on 14 July. The detachment arrived at Milne Bay on 18 July and the move was completed by the end of the month. However, the new airfield was badly waterlogged and still lacking dispersals and hardstands, so the aircraft were initially based at 30-Mile (Rorona), where they had arrived on 21 July after a ferry flight via Cairns and Seven-Mile.

❧

The discovery of the Japanese ships lying off Guadalcanal and Tulagi on 6 July encouraged more long-range reconnaissance of that area, with these missions also providing an opportunity to bomb the ships where possible. To counter these efforts, the newly arrived Yokohama *Ku* fighter detachment at Tulagi had around a dozen Nakajima A6M2-N Rufes under the leadership of *Buntaicho* Lieutenant Sato Ri'ichiro. While the "float Zeros" would struggle to catch a high-flying B-17, the altitude-limited LB-30s would prove more likely prey for the floatplane fighters.

On 9 July two 435th BS LB-30 Liberators flew from Townsville to Port Moresby for a reconnaissance and strike mission of the Tulagi area the following day. These were AL515 flown by Lieutenant Casper and AL573 flown by Lieutenant Wallace Fields (with Captain Dick Ezzard alongside as a check pilot). A destroyer was seen off Tulagi, while three transports were moored off Guadalcanal. These were attacked at 1200, but no hits were seen, with light AA fire noted coming from the shore.

The Liberators were then attacked by two Rufes flying a combat air patrol. As the Rufes climbed to intercept they were photographed by a crewman with a hand-held K-20 camera. Two Rufes were captured in the same frame, providing the first solid evidence of the new type to Allied intelligence. The Rufes fired 100 x 20mm and 800 x 7.7mm rounds, as against 11,000 rounds of defensive machine gun fire returned from the bombers.

The bombers were correctly identified as B-24s by the Rufe pilots, with one claimed as the unit's first kill. AL573 had indeed been hit with its Number 3 engine shot out. Also, the release mechanism of the auxiliary bomb bay fuel tank had jammed when the tank was only partially dropped. The return flight to Port Moresby was made on three engines and with the empty tank awkwardly protruding from the bottom of the fuselage. The following day, 11 July, both Liberators returned to Townsville.

On 12 July a B-17 scoured the same area and noted two floatplanes airborne, but no interception took place. This mission was repeated over the next two days later by other B-17s, by which time Allied intelligence was noting that an air patrol by two floatplanes was being maintained over Tulagi during daylight hours.

By this time Guadalcanal had been confirmed as a landing location for the US Marines in early August, alongside Tulagi. Elements of the 1st Marine Division had already arrived in New Zealand and there was an urgent need to gather intelligence about Guadalcanal as virtually nothing was known about the island. For this purpose two USMC officers, Lieutenant Colonel Merrill Twining and Major William Kean, were flown to Townsville from where they would ride as observers onboard a 435th BS B-17E fitted with cameras. After refuelling at Port Moresby, the Fortress (41-2639 piloted by Lieutenant Ramsey) proceeded to make several photo reconnaissance runs over the Tulagi-Guadalcanal area on 17 July.

During these runs three floatplanes were seen to take off far below. In fact, six Rufes were already aloft flying a combat air patrol, led again by the unit commander Lieutenant Sato Ri'ichiro. At 1340 the Rufes attacked the B-17, firing a total of 100 x 20mm and 650 x 7.7mm rounds, and claiming the bomber as shot down. The B-17 was undamaged but reported seeing a Rufe crash after descending in smoke and flames. This was FPO1c Hori Tatsuo who subsequently died of wounds sustained during combat or as a result of the crash landing.

The third LB-30 to join the 435th BS was AL570 named *Nipponese Nipper*, which only arrived in Townsville on 13 July. A few days later it was flown to Port Moresby by Captain Fred Eaton from where it would fly a strike mission against the Japanese airfield at Kieta on Bougainville. As noted earlier in this chapter, the Japanese had commenced and then abandoned airfield construction activities on this site. Eaton dropped 12 x 300-pound and 9 x 500-pound bombs on both the town and airfield, noting that the runway was obstructed by an X-shaped structure and that there was no activity in the town.

A Tainan Ku J1N1-C Irving reconnaissance twin, which made its operational debut on 20 July.

On 29 July 1942 Tainan Ku Zeros shot down five Douglas A-24s over the Buna area. This was the last time the American type was used in combat in New Guinea.

CHAPTER 5

BUNA LANDING
NEW GUINEA 21–31 JULY

As noted in Chapter 2, the IJA had been studying the possibility of an overland route to Port Moresby since June. This was given the codename Operation Ri, and for this reason several reconnaissance flights by Tainan *Ku* C5Ms over the Kokoda area were undertaken during late June and early July as already detailed. In relation to this study, plans were laid to land an advance detachment of the IJA's South Seas Force in the Buna-Gona area on 21 July. Subsequently it was decided to mount an overland offensive with the entire strength of the South Seas Force, so that the 21 July operation became the first of a series of landings instead.

The IJN saw this as an opportunity to establish a new airfield in northern Papua, so that a contingent of naval personnel would also be landed on 21 July. These were construction personnel of the 15th Establishment Unit (part strength), alongside 430 men from the 5th Sasebo SNLF (including detachments equipped with AA guns) and a communications unit. Subsequently the construction personnel would commence work on the airfield using local labour until further construction personnel and equipment were landed in early August.

A major impediment to the intended sea invasion of Port Moresby attempted in May was the slow convoy speed as some of the old IJA transports were only capable of 6½ knots. This meant a lengthy time at sea and exposure to attack from the air or by submarines. However, for Operation Ri three modern transports were chosen that were rated as "high-speed" (one of the ships, the *Kinryu Maru*, was capable of 18 knots). This meant that the route from Rabaul to Buna could be covered in less than 48 hours. In the eyes of Japanese planners this was a monumental strategic advantage.

The transports would be escorted by two ageing cruisers, three destroyers, a minelayer and three submarine chasers. Thirteen floatplanes from the *Kiyokawa Maru* seaplane detachment were assigned air defence and anti-submarine duties, as well as long-range patrols, commanded by *buntaicho* Lieutenant Yamada Masaji. To achieve these goals Yamada's detachment hosted four types of seaplanes which had changed their tail-code prefixes from "R" to "RI" the week prior: four Aichi E13A Jakes (tail-codes RI-1, -6, -8 & -9), one antiquated E7K1 Alf (tail-code RI-7), six F1M2 Petes (tail-codes RI-14 through -19) and two E8N2 Daves (tail-codes RI-23 and -24). The six Petes were the pride of the detachment, and were all brand-new aircraft, including some with barely ten hours flying time per airframe. Small detachments of Petes would also temporarily station themselves at locations nearby such as Girua and Salamaua from where they would conduct forward defensive patrols.

The destroyers, submarine chasers and E13A floatplanes provided for a reasonable anti-submarine capability. The minelayer, *Tsugaru*, was included because it had the best AA defence of any of the IJN ships then in the area: four Type 89 5-inch guns in two twin mountings and four Type 26 25mm guns, also in two twin mountings. Details of the convoy are as follows:

Operation Ri First Convoy

- *Tenryu* (4,350-ton light cruiser; built 1919; 4 x 5.5-inch guns; 6 x torpedoes)
- *Tatsuta* (4,350-ton light cruiser; built 1919; 4 x 5.5-inch guns; 6 x torpedoes)
- *Yuzuki* (1,772-ton destroyer; built 1927; 4 x 4.7-inch guns; 6 torpedoes)
- *Uzuki* (1,772-ton destroyer; built 1926; 4 x 4.7-inch guns; 6 torpedoes)
- *Asangi* (1,400-ton destroyer; built 1925; 4 x 4.7-inch guns; 6 x torpedoes)
- *Tsugaru* (4,400-ton mine-layer; built 1941; 4 x 5-inch guns; 4 x 25mm guns; 1 x E7K Alf floatplane; used as AA ship)
- *CH-28, CH-29, CH-30* (420-ton submarine chasers; built 1942; 1 x 3-inch gun)
- *Kinryu Maru* (9,310-ton armed merchant cruiser; built 1938; 4 x 6-inch guns)
- *Ayatosan Maru* (9,788-ton IJN transport; built 1941)
- *Ryoyo Maru* (5,974-ton IJA Transport; built 1931)
- *Kiyokawa Maru* (6,860-ton seaplane tender; built 1937; 2 x 5.9-inch guns; 4 x E13A Jake, 1 x E7K1 Alf, 6 x F1M2 Pete and 2 x E8N2 Dave floatplanes, some operating from shore bases).

After leaving Rabaul on 20 July, the convoy proceeded south-west of New Britain under the cover of darkness that night. At 0820 the following morning the larger warships were seen by a B-17 on reconnaissance, accurately counted as six in total representing the two cruisers, three destroyers and minelayer. The vessels were sailing on a westerly course at a reported 12 knots, although they were inaccurately reported as "two battleships and four destroyers". With no transports seen, the significance of these warships was initially unclear, and poor weather thwarted any further observations until very late in the day. However, the early sighting did enable a strike force of half a dozen B-26s to fly up to Port Moresby from Townsville.

There was little other air activity on 21 July, except that three *Kiyokawa Maru* F1M2 Pete floatplanes had forward deployed to Salamaua under the command of *buntaicho* Lieutenant Yamada Masaji. These were active during the day supporting friendly ground patrols inland from Salamaua and also over Buna. One of these, piloted by FPO3c Ogawa Yoshiji, was discovered by a patrol of nine 39th FS P-39s that afternoon after it was seen taking off. After an extended chase through clouds, it was claimed as shot down by Lieutenant Ralph Martin. However, Ogawa's Pete received only three bullet hits and he was able to land safely, before returning to the *Kiyokawa Maru* later that same day.

There were small parties of soldiers from the Papuan Infantry Battalion in the Buna-Gona area. These witnessed a floatplane machine gun the government station at Buna at 1440, and soon afterwards shipping was seen off the coast. The *Kinryu Maru* arrived three miles north of Buna late that afternoon and began disembarking the naval units. That evening the IJA advance force began disembarking from the *Ayatosan Maru* and *Ryoyo Maru* at Gona, a short distance west of Buna. No resistance was offered by a handful of Australian soldiers who reported the

SNLF soldiers wash their clothes while embarking on a voyage, New Guinea, early 1942. Marines of the 5th Sasebo SNLF were landed at Buna on 21 July.

developments and withdrew inland.

The news of the landing electrified the Allied Air Forces commanders in Port Moresby, who sought to arrange what attacks they could, although most of which would have to wait until dawn the following day. However, one B-17 noted the *Kinryu Maru* at Buna at 1700 and other ships offshore. A bombing attack was noted simply as "nil hits" and was recorded by the Japanese as targeting the *Uzuki*. Five of the B-26s attacked the shipping approaching Gona an hour later in conditions of poor visibility. A direct hit was claimed, but no damage was incurred. During these attacks the *Tsugaru* was busy protecting the transports, firing 108 rounds of 5-inch and 109 rounds of 25mm ammunition.

That night thousands of Japanese troops streamed ashore from the transports, and the bulk of the supplies were unloaded using landing barges. Three RAAF Catalinas had departed Cairns during the evening of 21 July to search the area that night, but despite fourteen-hour flights were unable to find the ships after reporting rain and 10/10 cloud in the Lae area. At 0600 on 22 July some of the warships bombarded the shore but all of the Australian soldiers in the area had already left. The morning of 22 July saw determined Allied attacks. These were aided by the fact that 18 Tainan *Ku* Lae-based Zeros were unable to fly their planned early defensive patrols above the ships due to inclement weather. Again and again, as we shall see, New Guinea's weather would not only govern aerial activity, but would often dictate its outcomes.

First to arrive that morning was a solitary No. 32 Squadron Hudson piloted by Flight Lieutenant Lloyd Manning which had taken off from Five-Mile 'drome (Wards) in darkness at 0300. In conditions of dim light, drizzling rain and a low cloud ceiling of less than 1,000-feet, Manning bombed an unloading transport at 0630 but reported the bombs falling 75 yards short. More than an hour later, at 0749, a 435th BS B-17E found the ships. The pilot, Lieutenant Robert M DeBord made a low approach, dropping 14 x 300-pound bombs in two passes. The first pattern of bombs narrowly missed a destroyer while the second pattern scored direct hits on the transport *Ayatosan Maru*. A fire started and the ship was subsequently beached and

Kiyokawa Maru Jake floatplanes on the shoreline of Malaguna Bay, Rabaul, several of which provided air cover for the first Buna convoy. The wartime censor has blurred the tail-codes.

disabled in shallow water. Henceforth it would become known to Allied airmen as the "Gona Wreck". The vessel had been mostly unloaded, but five soldiers and three crewmen were lost.

After the attack DeBord remained in the area for three hours, guiding further strikes and broadcasting regular reports through wet and overcast weather. Indeed, incoming from Port Moresby was a mix of B-25s, B-26s and P-39s. At 0855 six 13th and 90th BS B-25s bombed the *Ryoyo Maru* from 8,000-feet but scored no hits. These were followed five minutes later by five of six 2nd and 408th BS B-26 Marauders, with the sixth having aborted and returned to base. The B-26s unloaded 35 x 300-pound bombs but also missed.

The ground-attack Airacobras were a quartet of P-400s from the 40th FS and four more from the newly arrived 80th FS, covered at higher altitude by eight more Airacobras, four each from the 39th and 41st FS. Having ascertained there was no aerial enemy opposition, the American pilots turned their attention to ground targets. The 80th FS record states:

> These boys went right to work strafing a convoy off Buna and barges in the area, covering B-25s striking the same place. In fact they had a nice baptism of fire. On this mission or series they lost Lieutenant Hunter who apparently was hit by heavy ack ack and bailed out of his plane.

The attacking aircraft came under fire from the ships and also IJA AA positions setup overnight near Gona Mission. Lieutenant David "Pinky" Hunter was the first combat loss experienced by the 80th FS after being hit by Japanese groundfire. He was seen to jettison his door and ditch safely but came ashore not far from the unloading barges and was quickly captured. Like many captured Allied airmen he was subsequently taken to Rabaul where his beheaded corpse was exhumed post-war from a mass grave.

One of the four 40th FS pilots also failed to return and is believed to have been a victim of the bad weather. This was Lieutenant Garth Cottam who was last heard leading his element in an attacking dive through low cloud, before advising his wingman the weather was too thick to continue the operation. He most likely crashed into the nearby hills in conditions of zero visibility.

Next into the area were ten B-17Es of the 28th and 30th BS which had made the relatively unusual approach direct from the Australian mainland. These dropped their bombs against

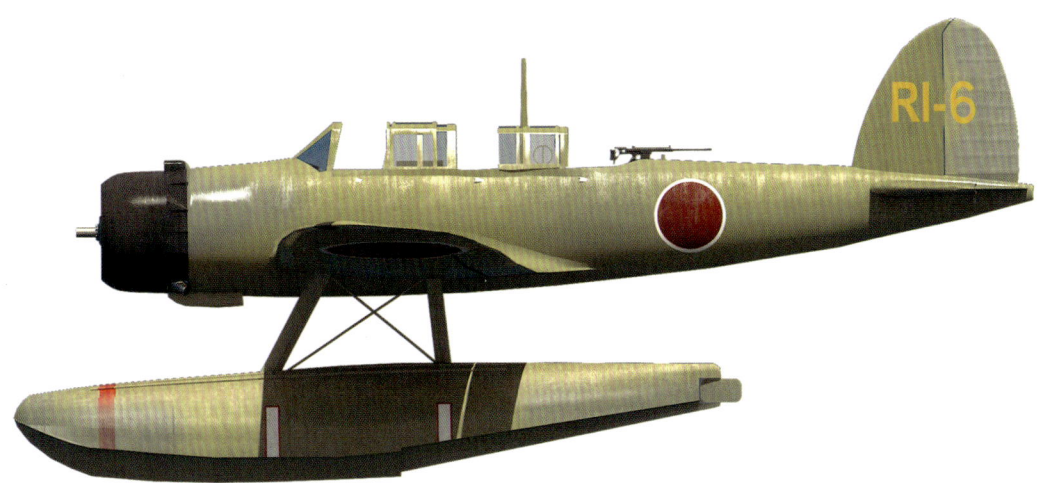

Aichi E13A1 Jake, CN 82, RI-6, Kiyokawa Maru, July 1942.

two destroyers and one transport. Major Rouse lamented:

> Found a perfect target and weather and no opposition but missed. The largest formation we have had and we missed.

Rouse's experience was just the most recent of many in the theatre whereby B-17s had missed moving ships when bombing from medium or high altitude, an inaccuracy which had not been fully appreciated before the start of the Pacific War. In contrast, DeBord's earlier success against the *Ayatosan Maru* had been from low altitude.

DeBord, meanwhile, had loitered in the area during the morning to monitor developments, and at 1050 had reported that the ships had left the beaches and were noted as out to sea. Port Moresby ordered other aircraft to reconnoitre the area and replace BeBord. The first of these was No. 32 Squadron Hudson, A16-173, piloted by Flying Officer Nathaniel Straus which departed Port Moresby at 1000. Arriving over the landing area in relatively clear weather an hour later, Straus made about ten runs over the area during which his crew operated an F-24 aerial camera. After Zeros were seen Straus left the area and on landing was met by a jeep which rushed the undeveloped film to Allied Air Forces headquarters. Straus had likely encountered six Zeros led by Lieutenant Kawai Zero which were the first to arrive at this time.

Prior to the return of Straus, a third No. 32 Squadron Hudson had departed Port Moresby at 1130. This was A16-201, flown by Pilot Officer Warren Cowan, who was tasked with relieving DeBord's B-17 and locating and shadowing the ships now known to be withdrawing. Two hours later it reported that it had passed over Gona and was about twenty miles out to sea. That was the last message received from the aircraft, but its fate is known from Japanese unit logs, elaborated by the memoirs of FPO1c Sakai Saburo.

Sakai was among six Zeros led by Lieutenant Sasai Jun'ichi which had left Lae early that

The wreck of the Ayatosan Maru, which was disabled off Gona on 22 July after being bombed by a 435th BS Flying Fortress.

afternoon to patrol over the Buna-Gona area. Almost immediately they spotted Cowan's lone Hudson, which went into a dive to gain speed in a bid to reach Milne Bay. The Zeros jettisoned their drop tanks and set off in pursuit, soon catching up to the Hudson and commencing stern attacks.

In such a situation, the position of the Hudson was usually not hopeless. An attacking Zero had to brave defensive fire from the turret's twin 0.303-inch machine guns, and after each attacking run the Zero would have to break off and spend time and fuel catching up to the Hudson again. However, against six-to-one odds Cowan did something radical and unexpected.

Cowan banked the Hudson steeply and stood one on wing, turning quickly to face his pursuers and firing the Hudson's twin forward facing 0.303-inch machine guns. At first the Japanese were taken aback by the audacious tactic and the unexpected manoeuvrability of the twin-engine bomber. However once among the gaggle of Zeros, Cowan's fate was sealed, especially after a burst from Sakai silenced the turret gunner. Quickly hit by further bursts, flames streamed out from the Hudson and it crashed into the jungle canopy at high speed. The wreckage was found in 1945 and described as:

> ... scattered for hundreds of feet throughout the thick jungle.

Meanwhile sixteen P-39s from the 39th, 40th, 41st and 80th Fighter Squadrons were in action over the beach strafing the barges and supplies unloaded earlier that afternoon. These were joined by six 2nd and 408th BS Marauders, with Brigadier-General Kenneth Walker riding as an observer in *Cossack*. During the course of the day the destroyer *Uzuki* received minor damage during the air attacks and sustained 16 casualties.

Also in action were seven No. 76 Squadron Kittyhawks. These were led by Squadron Leader Peter Turnbull, an experienced ace who had flown with No. 75 Squadron in April and before that in the Middle East in 1940-41. The Kittyhawks were loaded with 500-pound bombs with the instruction to dive-bomb the same AA positions near Gona that had earlier accounted for Hunter's Airacobra.

The Kittyhawks were escorted by eight Airacobras, but lost contact with them while crossing the mountains. Then, in clearer skies, at 1530 Turnbull's formation ran into Sasai's six Zeros, fresh from their victory over Cowan's Hudson. The Australians wisely jettisoned their bombs

Miss Snafu was an original 13ᵗʰ BS Mitchell, seen here in the dispersal area of Kila 'drome. This squadron's Mitchells were active targeting the Buna invasion convoy on the morning of 22 July.

and then fought a running battle at low level for some 50 miles as they retreated to Port Moresby. This was a heated engagement, as evidenced by the fact that the Zeros expended virtually all of their remaining ammunition. In a vindication of No. 76 Squadron's training and tactics, they experienced no losses and the only damage was a sole bullet hole in Turnbull's Kittyhawk A29-39 named *Abdul the Bull*. During this chase, which descended to low-level, several Kittyhawks ran perilously short of fuel, and three chose to break left and land at Milne Bay. Of the four which reached Port Moresby, one force-landed on the bank of a river not far from 30-Mile 'drome when it ran out of fuel, but it was later flown out.

The final Allied efforts against the beachhead occurred late in the afternoon after the Zeros of the final patrol were engaged with the Kittyhawks before returning to Lae. Five B-25s from the 13ᵗʰ and 90ᵗʰ Bombardment Squadrons bombed barges at Buna from medium altitude, followed by a single 435ᵗʰ BS B-17E. The last sortie of the day was by four 19ᵗʰ BG Fortresses from Horn Island, which in the absence of other targets bombed the already stricken *Ayatosan Maru*.

During the course of the day the Japanese had recorded attacks by approximately 100 aircraft attacking in six or seven waves. Aside from the loss of *Ayatosan Maru*, relatively little damage was inflicted, largely because the ships had mostly been unloaded overnight. Colonel Charles Willoughby, who was MacArthur's chief of intelligence, despairingly wrote:

> For the record ... in spite of our advance notice, our air force was unable to prevent this landing.

Such an assessment needs to be balanced by the challenges imposed by the bad weather conditions, which allowed the ships to reach the coast safely after only fleeting attacks late on

B-26 Cossack force-landed at Seven-Mile after receiving combat damage over Buna on 22 July 1942.

Nakajima E8N2 Dave, CN 565, RI-24, Kiyokawa Maru, July 1942.

21 July. Nevertheless, the outcome did not heighten MacArthur's view of the efficiency of the Allied Air Forces at this critical time.

Operations on the following day, 23 July, were again limited in effectiveness due to more poor weather. Various Hudson and B-17 reconnaissance missions were flown, but little was seen, although a Hudson dropped bombs over Buna just after dawn. In response to the *Kiyokawa Maru* floatplane operations at Salamaua (witnessed by Australian soldiers in the nearby mountains) and the sighting of a destroyer there, believed to be the *Yuzuki*, four 40th FS Aircobras were sent on a dawn strafing mission but results were "not observed" amid the low cloud and mist.

Five main strikes now proceeded against the Buna-Gona area. The first of these comprised thirteen Airacobras, of which only five persisted to strafe huts at Gona Mission around 0900. These were followed two hours later by five A-24 dive-bombers which were among eight that had been effectively retired by 8th BS but, in a desperate measure, had been hurriedly flown up to Port Moresby from Charters Towers the previous day. The A-24s bombed barges and buildings at Gona, noting no activity except some AA fire. The AA positions were strafed by four P-400s which accompanied a single B-26 that bombed a ship off Buna, likely to have been the *Ayatosan Maru* wreck.

These attacks were followed by seven B-26s at 1115 which targeted "enemy stores and installations" at Gona. However due to the heavy overcast three of the aircraft returned without dropping their bombs. After a possible break in the weather late that afternoon, the seven B-26s returned to the area at 1600. They targeted wooded areas half a mile behind the Buna shoreline with "results unobserved". In actual fact such attacks accomplished very little, with most of the IJA troops already pushing inland.

The day did feature an unusual encounter further afield that afternoon. A C5M2 Babs piloted by FPO2c Hanahiro Keiryu was airborne over Rabaul when a lone 435th BS B-17E was sighted. The Babs dropped two 30-kilogram aerial bombs on the intruder at 1315, but no damage was inflicted. The Fortress misidentified the Babs as a Kate torpedo bomber. The aerial bombs were a creative weapon which theoretically showed promise, having a claimed blast radius of 75 yards. They were occasionally used but, despite generous claims to the contrary, achieved no confirmed success over Rabaul.

Meanwhile, over at Vunakanau that morning 23 No. 4 *Ku* Bettys led by Lieutenant Egawa Renpei had departed at 0815. They struck capricious weather almost immediately and Egawa gingerly turned the formation for home, all landing safely at 1005.

Slightly improved weather on 24 July saw an increase of air activity across New Guinea, although cloud persisted in most areas. The pattern of Hudsons departing Port Moresby in darkness to check the Buna-Gona area at dawn continued, this time being Sergeant "Nobby" Clarke in A16-192 which took the opportunity to drop bombs on Gona Mission at 0630. Further attacks against the same target area soon followed, with eight A-24s bombing AA positions at 0845. The dive-bombers were escorted by eight 39th FS P-400s. On this occasion coordination between the Banshees and Airacobras was very good, with escorting pilot Lieutenant Charles King noting in his diary that it was a:

... model of precision attack and formation flying.

Ten minutes later, six B-26s bombed barges on a beach near Gona. The Marauders were escorted by four P-400s but on this morning no Zeros were present. AA fire was reported as "intense, light, inaccurate". Some hours later, at 1400, seven B-26s were back and bombed buildings at Gona Mission.

From Rabaul the same 23 No. 4 *Ku* Bettys which Lieutenant Egawa Renpei had tried to get through to Port Moresby were back in the air, this time escorted by fifteen Tainan *Ku* Zeros. Egawa, spooked by the encounter with bad weather the previous day, refuelled his fleet at Lae on the way to and from the target. As these approached Port Moresby three bombers aborted for unstated reasons, one returning to Lae and two to Gasmata. This time the target was Twelve-Mile (Bomana) and the Australian Army recorded "about 150 anti-personnel bombs" dropped within the Twelve-Mile camp area, underestimating the actual bomb drop of 240 x 60-kilogram bombs. Some dumps containing fuel and bombs were destroyed but there were no casualties. Sixteen Airacobras scrambled but due to a lack of warning and cloudy conditions no contact was made with the enemy aircraft.

Meanwhile army headquarters in Port Moresby now faced threats on two fronts. On the first, from Lae and Salamaua Japanese patrols were aggressively pushing against the forward-deployed units of Kanga Force supplied from Wau. On the second front, from Gona the freshly landed advance units of the South Seas Force were pushing inland towards Kokoda. With both Wau and Kokoda only connected to Port Moresby by narrow mountain foot tracks that took the fittest more than a week to traverse, the land campaign began to take a new angle. Supply of forward deployed units via these tracks was remarkably inefficient as it depended on human "carriers", and soon both sides would deploy large amounts of natives in this capacity, including some of those used by the Japanese imported from Rabaul. However, the longer the distance to be covered the amount of supplies that could be carried per head declined markedly as the carriers also had to transport food and supplies for themselves.

The obvious alternative to supply via the foot tracks was air transport. However, as already outlined air transport resources in the SWPA were sparse and already overworked. The best that could be brought up at short notice was a couple of C-53s, an early paratrooper version of the original Douglas DC-3, the first of which flew into Port Moresby on 24 July and was immediately used to fly supplies to Wau. At the same time army headquarters wanted a signal detachment of the 39th Infantry Battalion and the battalion's commanding officer, Lieutenant-Colonel William Owen (a veteran of the disastrous defence of Rabaul in January), flown into Kokoda to join a forward company which had arrived there by foot. To accommodate this Hudsons of No. 32 Squadron were used in a transport capacity. The first flight on 23 July was forced back due to weather. Flight Lieutenant John Murphy made a fresh attempt the following day:

We found out that there was an old, unused strip at Kokoda ... but information on its length,

This is the only known photo taken of 3ʳᵈ BG A-24s departing for a combat mission, taken at Fourteen-Mile in mid-1942. By July the dive-bombers had been effectively retired but eight were rushed back to Port Moresby on 22 July following the Buna landing.

A Betty bomber launches from Vunakanau. On 25 July ten Bettys were forward deployed to Lae from where they flew a series of intimidation missions against nearby villages.

The damage incurred on 27 July on the ground at Horn Island when B-17E Flying Dutchman collided with Tojo's Physic.

serviceability etc was non-existent. After climbing to 10,000 feet approx. to get over "The Gap", the strip was sighted and after a couple of passes a landing affected. Kunai grass about three-foot long helped the landing but made the take-off a virtual impossibility. Luckily, we were able to recruit a very large number of local natives who, with great gusto and much laughter and noise, virtually cleared a take-off strip. After the natives pushed our aircraft backwards to get maximum take-off length, running up to full take-off power on brakes and after a short prayer we were away.

Weather again frustrated transport operations on 25 July, but the next day two C-53 flights enabled a full platoon of D Company, 39th Infantry Battalion, to be flown into Kokoda. That evening, with enemy troops pressing towards the area, Owen sent a message that Kokoda would likely fall to the enemy and it was too dangerous for further flights to be made. However, before the message was passed on another two flights of reinforcements had departed Port Moresby the following morning. These circled Kokoda around midday where the enemy had not yet appeared and those on the ground hurriedly tried to remove obstructions placed across the runway. However, the transports then received orders recalling them. As they flew away to the south those on the ground felt intense disenchantment at having been abandoned.

Meanwhile the aircraft transport *Mogamigawa Maru*, escorted by the destroyer *Akikaze*, had departed Rabaul on 23 July and arrived at Lae on 25 July. Among its unloaded supplies were critical aircraft spares for the Tainan *Ku*, and the ship departed later the same day. These vessels had been spotted by an Allied reconnaissance flight at 1423 on 23 July while north of New Britain. Normally such a detection would have provoked a frenetic response but given the poor weather and the Allied focus on the Buna-Gona beachhead, the voyage was undisturbed.

Air operations on 25 July opened with a night mission by a Catalina which had departed Cairns late the previous afternoon and which bombed targets on Gona beach in the early hours. Also in the air was a single B-25 which dropped bombs on Buna at 0053 by dead reckoning due to bad weather, an exercise of hope over futility. This was followed at dawn by another single B-25 which bombed the Gona Mission area.

A formation of eight 13th and 90th BS B-25s was next over the area at 0920, escorted by six P-39s, and all quickly found themselves fending off Zeros as for the first time in three days the weather at Lae had enabled them to take-off early. This was a nine-strong *chutai* led by Lieutenant Sasai Jun'ichi, whose pilots had previously accounted for Cowan's Hudson. The Zeros forced the B-25s to jettison their bombs in trees 400 yards south of Gona Mission, and then made repeated attacks on the bombers, six of which were damaged to some extent.

As combat between the P-39s and the Zeros developed, also running into the area were six 40th FS P-400s each hauling a 300-pound bomb and tasked with attacking the Gona-Kokoda "road". Lieutenant Frank Beeson was missing in action after this encounter. The burned-out wreckage of his Airacobra was discovered near Buna in April 1943 with the 300-pound bomb still attached, indicating he had been taken by surprise. Beeson's remains were found nearby with a single round missing from his sidearm, suggesting the wounded flyer did not want to be captured alive.

Also hit was Lieutenant David Hoyer who, despite suffering severe blood loss and fractures in his legs, was able to nurse his damaged fighter back to Port Moresby. Hoyer needed to be lifted from his cockpit and recovered after extensive convalescence, but his aircraft was written off.

That afternoon 435th BS B-17E *Gypsy Rose*, flown by Captain Maurice Horgan, was flying a reconnaissance of Rabaul and Kavieng. However due to persistent overcast he had to descend to just 4,000-feet and was intercepted by a nine strong *chutai* of Tainan *Ku* Zeros led by *buntaicho* Lieutenant Kawai Shiro, arguably the most aggressive pilot in the unit. These Zeros made persistent and repeated attacks on the Fortress over a 45-minute period which only ended at 1415 when the bomber escaped into cloud. During the attacks the Fortress was holed many times, with one engine put out of action and another damaged. In addition, two gunners were wounded, but Horgan was able to return to Port Moresby where his crew claimed six Zeros destroyed. Likewise, Kawai's pilots claimed the Fortress, but in fact neither side suffered a loss.

Later a force of six B-25s was over Gona Mission at 1720, the third appearance of Mitchells there on this day. These dropped 45 x 300-pound bombs on suspected barge positions from 10,000-feet, although no results were observed.

Another curious series of missions flown on 25 July was by ten No. 4 *Ku* Bettys which had arrived at Lae. Throughout the day they flew four missions in trios and pairs where they dropped 84 x 60-kilogram bombs and fired 2,900 rounds of 7.7mm ammunition from low altitude. For three of the missions the target was Gabmatzung Mission and for the other it was Mubo. Both of these locations were only a very short flight from Lae but the missions typically lasted two to three hours suggesting an element of surveillance was also undertaken. These areas were in the vicinity of Kanga Force forward camps, but while the Australian Army noted a "raid" at Nadzab (adjacent to Gabmatzung Mission) on this date it seems no Australians were directly targeted. Instead, it is likely that the Japanese saw the missions as serving the purpose of intimidating the populations of local villages, some of which villages were recorded as being burnt by the Japanese at this time.

The date of 26 July saw a disastrous mission for the 3rd BG. Five B-25s had flown into Port Moresby the previous evening from Charters Towers, and these were tasked with attacking Gasmata on New Britain where it was erroneously believed that the flying boats that had bombed Townsville overnight had alighted (see Chapter 7 for details). However, the flightpath took the Mitchells close to the Buna-Gona beachhead over which Sasai's *chutai* of nine Zeros was patrolling, having launched from Lae early that morning.

The Mitchells were to be escorted by six 40th FS P-39s, but the join-up had been tardy and the fighters lagged behind the bombers which were cruising at 14,000-feet. Near Buna the B-25s were attacked by the Zeros and responded by jettisoning their bombs and turning for home. However, two B-25s were disabled on the first pass and were soon plummeting earthwards in flames. From one of these aircraft, named *Aurora*, pilot Captain Frank Bender and one other crewman were able to parachute safely. Bender was able to find the wreck site with the help of natives, and buried the remains he found:

The gruesome part of this I won't go into in detail; it will suffice to say that I buried my engineer and bombardier in their parachutes. I believe these brave men were killed by the cannon shell that wounded me, for it must have hit on the nose of the ship. I also saw my co-pilot, who evidently had been overcome by smoke when he first left the co-pilot's seat and never did get out. The two rear gunners had bailed out and landed somewhere, but I couldn't find them. The part of the ship that I found was so well burned that I could salvage nothing, and the rear end of the ship, where the box of rations was kept, was nowhere around so it must have been blown off and away from the rest of the ship in the explosion.

Bender was soon joined by the other crewman who had parachuted safely, a member of the RAAF leant to the Americans as a turret gunner, Technical Sergeant Arnold Thompson. Both men spent weeks in the mountains before eventually returning to Port Moresby. Shortly after the war, Bender's experience was dramatised in a *True Comics* war action cartoon named *Hit the Silk*.

Meanwhile the other three B-25s were pursued all the way back to Port Moresby by some tenacious Tainan *Ku* pilots. All three B-25s were badly damaged, and FPO1c Sakai Saburo topped off the occasion by claiming to strafe RAAF Kittyhawks parked at Fourteen-Mile 'drome (Laloki). This incident was dryly recorded by the Australian Army as:

A single Zero, which evidently followed in two of our returning B-25s was engaged by MMGs [Medium Machine Guns] over Laloki drome 0920-0930 ...

All of Sasai's Zeros returned jubilantly to Lae, where they claimed three B-25s destroyed, only a slight overestimate of the two shot down and three damaged.

Later that day, shortly before sunset at 1745 a trio of Marauders was searching for the transport route between the Buna-Gona beachhead and the Wairopi Bridge. Failing to find it amidst thick cloud, one B-26 lost touch with the others and returned to base with its bombs still onboard. However, the other two B-26s came upon a destroyer lying one mile offshore, against which they dropped 50 x 100-pound bombs from 3,000-feet. The Americans claimed no hits while noting slight medium but inaccurate AA fire.

In fact the destroyer, the *Asanagi*, was forced to manoeuvre hard to avoid the bombs during which the ship incurred light damage after scraping its hull on a reef. This was a close escape for the destroyer that subsequently underwent emergency repairs at Truk before returning to Japan. The *Asanagi* was in company with an unidentified transport ship which had successfully run the second supply convoy from Rabaul to Buna. Evidence of casualties during this attack is given by the fact that an IJA staff officer named Tsuji had departed Rabaul for Buna on a destroyer on 25 July. Ultimately, he did not disembark as he suffered a throat injury after being struck by shrapnel and was repatriated back to Rabaul the following day.

The day concluded with two Catalinas loitering over Gona's coastal area that evening, dropping a mix of 500-pound general purpose and 4- and 20-pound incendiary bombs. These "nuisance" raids continued nightly until the end of the month.

The two-day period of 25-26 July also saw a remarkable pair of missions by a hard-worked LB-

A 19ᵗʰ BG Fortress being bombed up under a camouflage net at Mareeba, 1942. Use of this location from late July greatly facilitated operations as it was 500 miles closer to New Guinea than the previous base at Longreach.

30 crew which saw them clock up 23 hours of flying with only three hours of sleep. This was AL570 flown by Captain Fred Eaton of the 435ᵗʰ BS and his RAAF co-pilot Sergeant Mervyn Bell. The Liberator departed Townsville at 1200 on 25 July for an armed reconnaissance of Buka, where the Japanese had been considering upgrading the former RAAF airstrip and some activity had been reported. However, the crew found the area covered in thick cloud. After spending more than an hour overhead they bombed through the cloud from 6,000-feet and could not observe results.

After arrival at Port Moresby shortly before midnight, the crew were informed of an early morning mission and were awoken at 0430. However, that morning priority was given to fuelling and arming the B-25s, B-26s and P-39s for the missions noted above which unfolded on 26 July. Subsequently the Liberator was not airborne until 0900, after which it quickly landed again due to some technical problems. The bomber was finally off on its mission at 1200, returning to Bougainville where it bombed transports in the harbour but recorded no hits. The Liberator returned to Townsville later that evening.

The 27 July saw a relative lull in operations due to overcast and rainy conditions. At 1020 three B-26s escorted by Airacobras managed to bomb Gona Mission. Three hours later five A-24s were sent to bomb a ship reported off Gona, escorted by twelve Airacobras. The ship was not seen and instead the dive-bombers targeted the road leading away from the beachhead area, believing they may have hit a small ammunition dump after witnessing an abnormally large explosion.

After failing to land troops at Kokoda on this day, the two C-53s were sent to drop supplies there.

They were to be escorted by eleven P-400s which would then strafe the Buna-Kokoda road. However due to encroaching weather neither operation was carried out. Japanese operations were also curtailed on this day, although a C5M Babs managed a reconnaissance mission over Buna, Kokoda and Port Moresby during the afternoon.

Meanwhile nine Fortresses led by Major Dean "Pinky" Hoevet, the CO of the 30th BS, had also tried to bomb Buna but due to the thick overcast his bombers turned back for Horn Island. While coming into land, the B-17E 41-2460 *Flying Dutchman* flown by Lieutenant Edward Bechtold slammed into another B-17E on the ground waiting to take-off. The latter bomber was 41-2640 *Tojo's Physic* flown by Captain Carey O'Bryan. Major Rouse commented:

> Becktold [sic] ground looped on landing and tore the nose completely off one of our ships and smashed his up pretty badly. Purely pilot error due to inexperience, I believe. Two more ships gone that we needed badly.

Bechtold's B-17 was immediately written off while *Tojo's Physic* was initially flown to Mareeba for repair, however it was subsequently decided that it couldn't be used for combat and it was used as a spares source.

The next day, 28 July, saw bad weather again greatly limit operations. Three B-26s attempted to attack Lae but could not approach the target area due to cloud cover. Two additional transports flew into Port Moresby from Queensland, a DC-2 and a Lodestar. These combined with the C-53s to make attempted supply drops to Wau. Two drops were completed but another three trips were unsuccessful due to bad weather. Also on this day a sole F-4 Lightning reconnaissance aircraft of the 8th PRS arrived at Port Moresby and was tasked with reconnaissance of the Buna-Gona area, weather permitting.

Late that afternoon a trio of No. 4 *Ku* Bettys departed Vunakanau at 1630 in an attempt to raid Port Moresby. They ran into bad weather and spent half an hour trying to locate their target that evening, before dropping 27 x 60-kilogram and 9 x 70-kilogram bombs – the latter size being unusual and rarely appearing in Japanese records. The raid was recorded from the ground as being "probably one 4-engined flying boat" which scattered bombs between Three-Mile and Seven-Mile that did no damage. AA guns opened fire but made no observations due to heavy cloud, although flares were reported which might explain the unusual 70-kilogram ordnance.

The 29 July initially unfolded quietly, with no offensive operations by either side. However shortly after 1400 ships were sighted approaching Gona. This was the third Buna convoy, consisting of these five ships:

Third Buna Convoy

- *Tatsuta* (4,350-ton light cruiser; built 1919; 4 x 5.5-inch guns; 6 x torpedoes)
- *Yuzuki* (1,772-ton destroyer; built 1927; 4 x 4.7-inch guns; 6 torpedoes)
- *Kotoku Maru* (6,702-ton IJN ammunition carrier)
- *Ryoyo Maru* (5,974-ton IJA Transport; built 1931)
- 1 x Submarine Chaser – unidentified likely CH-29 or CH-30

By mid-afternoon the skies had cleared sufficiently for operations and eight 8th BS A-24s departed Port Moresby with a heavy escort of twenty Airacobras, ten each from the 40th and 80th Fighter Squadrons. The dive-bombers were led by the 8th BS commanding officer, Major Floyd "Buck" Rogers, who was no doubt keen to restore the reputation of the Allied Air Forces and strike a telling blow against the enemy.

However, while over the Owen Stanley Ranges things began to go awry. First, one A-24 turned back to base after experiencing engine trouble. Then the respective A-24 and P-39 pilots experienced difficulty keeping in contact given their different cruise speeds and enforced radio silence. By the time that Rogers' men were approaching the target area they had separated from their escorts altogether. Then bad turbulence shook the bomb loose from underneath Lieutenant Robert Cassels's A-24, but limited to using hand signals, his fellow pilots were unable to communicate what had happened. Hence Cassels flew on, albeit confused as to what his comrades were trying to tell him.

As the A-24s reached the push-over points for their dives on the ships below at 1600, they were in conditions of good visibility. Indeed, the weather on this side of the ranges had been good enough for Zeros to fly defensive patrols over the beachhead area since early that morning. The fifth and final patrol was now in the area comprising a full nine-strong *chutai* led by Tainan *Ku buntaicho* Lieutenant Yamashita Joji. Two of Yamashita's highly experienced pilots had been involved in shooting down two 3rd BG B-25s in the same area four days earlier.

Rogers completed his dive, targeting a transport which he missed by some 50 yards, and was then set upon by two Zeros. Pursued at sea level, Rogers was quickly shot down and his stricken dive-bomber ploughed into the sea. Following Rogers was Lieutenant John Hill, who also missed the ship and then found himself also pursued by Zeros at low level. His rear gunner, Sergeant Ralph Sam, was badly hit and when he couldn't man his twin 0.30-inch calibre machine guns he fired his sidearm. Hill subsequently made an emergency detour to the newly completed airfield at Milne Bay where he got Sam to a rudimentary field hospital, although he later died of wounds after having been evacuated to Australia.

At Port Moresby, first news of the unfolding disaster was the return of Lieutenant Raymond Wilkins, who had accompanied Hill back from Milne Bay. Wilkins and Hill were the only two pilots to return from this mission, making five of seven A-24s unaccounted for. Subsequently it was found out that Captain Virgil Schwab's dive-bomber had been hit by AA fire and crashed on a beach north of Buna in a ball of fire killing both crew. Three other A-24s were disabled with the crews either effecting forced landings or bailing out. All six men were known to have survived before being caught by Japanese forces and were among a party of European prisoners that was executed at Buna on 12 August by Lieutenant Komai Uichi of the 5th Sasebo SNLF.

The disaster meant that when all surviving A-24s returned to Australia they were ordered never to be used in combat again. The Japanese pilots incorrectly but understandably identified their victims as naval SBD Dauntlesses, and hence briefly feared an American carrier force was nearby. One of the A-24 pilots had managed to hit the *Kotoku Maru*, with the ship taking on water and listing heavily to starboard. Some 263 IJA engineers aboard were safely put ashore in motor

The wreckage of one of the A-24s shot down on 29 July as it lies in jungle today.

launches but the cargo was unable to be unloaded. That night the *Kotoku Maru* made it to Lae under reduced power and effected some temporary repairs before heading back to Buna in an attempt to unload the cargo the following day.

Other Allied attacks to strike the ships late that day were unsuccessful, impeded due to failing light and weather. Of two B-17s one jettisoned its bombs after a technical failure and the other returned with its bombs after failing to find the target. Three Hudsons couldn't find the target either and instead dropped their bombs near the Gona Wreck. During the afternoon four transports dropped supplies at Wau and a single transport did the same at Kokoda.

The evening of 29 July saw a nuisance raid on Port Moresby, with three Bettys making three separate runs over the target area during a two-hour period commencing 1853. Once again, the No. 4 *Ku* pilots had some difficulty locating targets in bad weather. Some 25 x 60-kilogram bombs were dropped, landing near Seven Mile and searchlight positions, but doing no damage.

Several hours later the mission was repeated by another trio of Bettys, which made separate runs during a half hour period commencing 0245 on 30 July. Searchlights illuminated the planes which were fired on by AA batteries, but all of the bombers returned to Vunakanau safely. The bombs fell near a corner of one of the Seven-Mile runways but did no damage. A third trio of Bettys which had departed Vunakanau half an hour after the second formation later dropped 34 x 60-kilogram bombs but they too were unsure if they were over the target area – this last raid was unreported at Port Moresby so likely the bombs fell harmlessly into the jungle or the ocean.

The morning of 30 July saw bad weather once again limit operations, with a B-17 reporting complete overcast at Lae and an F-4 Lightning unable to photograph Buna for the same reason. With the situation at Kokoda becoming dire, three transports and a Hudson managed to drop supplies to Australian troops in the mountains. Earlier three No. 32 Squadron Hudsons had flown a bombing mission against Buna, but that was the last such mission flown by the squadron as the Hudsons were increasingly used for supply drops.

Otherwise the only action that morning concerned a lone 435th BS Fortress which reported sighting three destroyers and one transport in waters north of Papua and heading north at 0835. This was the *Ryoyo Maru* which had left the beachhead area early that morning for the return voyage to Rabaul. Providing cover for the transport were nine Zeros of the Sasai *chutai* which made repeat attacks on the bomber during a period of some 90 minutes. The Tainan *Ku* pilots expended some 3,000 rounds of ordnance, but the Fortress was able to hide in cloud cover for much of the engagement. It was claimed as destroyed but there were no losses to either side.

With the exception of the 435th BS, which was based at Townsville, the other three squadrons of the 19th BG had been operating from the outback Queensland airfield at Longreach for some months. By 24 July they had moved to Mareeba, which was some 500 miles closer to New Guinea and meant much greater flexibility. Rather than offering a "next day" service and overnighting at Port Moresby, the 19th BG could now theoretically attack targets around Papua directly from Mareeba and refuel at Port Moresby or Horn Island on the way home.

Hence after the ship sighting that morning two formations of Fortresses were sent out to strike shipping in the area, arriving very late that afternoon. The first group of eight B-17s arrived over the beachhead area at 1715 with no shipping in sight save for the burning Gona Wreck. This was attacked by some of the Fortresses, recording two hits, but several of the aircraft continued flying along the coast to search for targets. The persistence was rewarded 45 minutes later when they came upon the *Kotoku Maru* near Salamaua while attempting to return to Buna from Lae.

The Americans claimed five direct hits on the transport, likely bombing from low to medium altitude as they also reported strafing the transport and a destroyer from "200 to 1,500 feet". The *Kotoku Maru* was hit by at least three 500-pound bombs and was severely damaged, forcing the captain to run the ship aground to prevent it from sinking. Those aboard were rescued by the light cruiser *Tatsuta* and the destroyer *Yuzuki*, and the cargo was eventually salvaged although the ship remained as a wreck. The second formation of six B-17s from Mareeba was unable to locate any targets.

There were few air operations on the final day of the month, 31 July. Both sides were preoccupied by the news that the IJA advanced force had captured Kokoda in the early hours of 29 July during a night action where the Australian commander, Lieutenant-Colonel William Owen, had been killed.

B-17E Jap Happy departs Efate in the New Hebrides on 31 July 1942 as part of the 11th BG's first mission against Guadalcanal.

CHAPTER 6

TARGET TOWNSVILLE
SOLOMONS & NORTH QUEENSLAND
21 JULY–1 AUGUST

A new entrant to the South Pacific theatre at this critical juncture in late July was the 11th Bombardment Group. This unit had been operating B-17Es since 1941 from Hickam Field, Hawaii, and had participated in the Midway operations in June. On 17 July the first of its bombers departed Hawaii for a trans-Pacific flight and arrived at Plaine des Gaiacs, New Caledonia, four days later. Within a few days the Group's 42nd and 98th Bombardment Squadrons had also arrived in the French territory, while on 25 July the 26th BS arrived at Efate in the New Hebrides. The Group's fourth squadron, the 431st BS, moved to Nadi, Fiji, at this time and did not participate in the initial operations described below.

For the present time these three squadrons fell under USN command, with Tulagi and Guadalcanal being priority targets for both reconnaissance and bombing. However, the airfields in New Caledonia and the New Hebrides had virtually no facilities except for dumps of fuel and bombs. Crews often slept on cots under the wings of the bombers, with the flaps lowered and bomb bay doors open to give protection from elements. Pilots took off from soft and often water-covered runways. On night take-offs, the only lights were rags stuck in bottles of gasoline and the headlights of jeeps at the ends of the runways. There were no navigation aids except ADF (Automatic Direction Finder) homing signals, which were reliable only at relatively short distances. They had the additional danger of the instrument's needle pointing directly at tropical thunderstorms due to static generated by lightning, giving a false reading away from the destination.

Nevertheless, the need for photo reconnaissance over Tulagi and Guadalcanal was so acute that only two days after arriving in New Caledonia three 98th BS B-17Es were tasked with the first mission. The Navy provided the cameras and the Marines the photographers (from VMO-251), and the mission departed from Plaine des Gaiacs on 23 July. During their photo runs the Fortresses were intercepted by a flight of Yokohama *Ku* Rufes from which Sea1c Matsui Saburo was shot down and killed.

At this time regular reconnaissances of Tulagi and Guadalcanal were also being flown from the SWPA, including a solo B-17 flight on 24 July which noted "the large 'drome well cleared towards completion". Three floatplanes climbed to intercept, but the B-17 took cover in cloud and no combat eventuated.

On 29 July the first B-17 landed at the newly completed airfield on Espirito Santo, which was 125 miles closer to Guadalcanal than Efate. This would serve as a forward field to enable missions to be flown against Guadalcanal, but was still very rudimentary as described by the

11th BG historian:

> A narrow strip cut partly from a coconut grove, partly from the encroaching jungle; revetments barely deep enough to keep a B-17s nose off the runway and so narrow a man had to stand at each wing tip to guide the pilots out to the short taxiway.

On 31 July the 11th BG launched its first bombing raid against Guadalcanal. This comprised six B-17Es from the 26th BS and two from the 98th BS, all chosen because they had been modified in Hawaii with additional fuel tanks installed in the forward fuselages, just behind the radio compartment. Because of this extra fuel, the Fortresses were limited to 2,000 pounds of bombs, in either 4 x 500-pound or 20 x 100-pound bomb configurations. After making the over-ocean flight at extremely low altitude because of a "murderous overcast", the B-17s climbed to 18,000-feet and circled Lunga Point to bomb the almost completed airfield. The 11th BG commanding officer Colonel LaVerne "Blondie" Saunders recorded that:

> Ack-ack was heavy but we got our bombs away on the target and returned to base undamaged.

In fact, Major James Edmundson, flying B-17E 41-2610, noted attempted interception by Rufes:

> We were flying a three-ship formation. Our B-17s were at 16,000 feet. The Japanese, using float-type Zeros, came in about 1,000 feet above us and to the right side of us. They didn't make any attempt at concealment. They rode alongside us and ahead 45 degrees off our nose for about five minutes.

It appears the Rufes respected the overwhelming firepower of a Fortress formation, and thus chose not to tangle with them. The pilots were likely mindful also of the loss of Sea1c Matsui Saburo a week earlier, although the Yokohama *Ku* log optimistically records that three Rufes "pursued" seven Fortresses, which escaped. With the American landing on Guadalcanal now just a week away, the 11th BG would repeat these efforts on a daily basis, weather permitting. It was critical that the Japanese airfield construction effort be delayed to prevent enemy aircraft from being based there that might contest the landings.

That same day three RAAF Catalinas were tasked with attacking stores dumps on Guadalcanal and ascertaining if the floatplanes were still based at Tulagi. For the latter purpose two Catalinas departed Havannah Harbour in Efate at 0700 and 0810 and arrived over the target area in daylight. Three Rufes were on patrol throughout the day led by Lieutenant Hoshino Hirokatsu who reported an attempted interception against a flying boat against which they fired a modest 15 x 20mm and 100 x 7.7mm rounds. The flying boat in question was Catalina A24-24 flown by Pilot Officer CW Miller who reported being pursued by two fighters "but not attacked" – evidently, Miller's crew did not notice the distant gunfire.

The two Catalinas noted ships lying off Guadalcanal but no flying boats moored at Tulagi as all of the H6K Mavises were away flying patrols. The third Catalina departed Efate that afternoon on a night mission, by which time it was able to observe several returned flying boats moored at Tulagi.

Fortresses of the 11ᵗʰ BG lined up at the newly built airfield on Espiritu Santo, 1942.

Airmen shackle 100-pound bombs for an 11ᵗʰ BG B-17 mission, Espiritu Santo, 1942. A typical load for the early missions to Guadalcanal was 20 x 100-pound bombs.

Kawanishi H6K Mavis, Y-44, Yokohama Ku, Tulagi, July 1942.

The lumbering but majestic H8K series Emily flying boat was the largest Japanese aircraft to grace South Pacific skies. The first to visit Rabaul were a pair of Yokohama *Ku* Emilys which arrived in March. However as recounted in *South Pacific Air War Volume 2* their operations in the area were only brief. After one of the flying boats vanished during a routine patrol the other returned to Japan.

More than three months later in July 1942, a detachment of four H8K1 Emilys serving with the No. 14 *Ku* arrived at Rabaul from Jaluit in the Marshall Islands, sporting tail codes W-37, W-45, W-46 & W-47. They were the only aircraft with sufficient range to bomb Townsville, a long-held ambition of Japanese strategic planners. On 25 July 1942 W-45 and W-46 departed Rabaul shortly after 1600 in the late afternoon and reached Townsville at 2330 that night.

The flying boat crews reported that bright lights lit the jetty, and that other lights were also visible in the town. It was a good half hour before the Australian military authorities in Townsville realised the aircraft might be hostile, by which time four searchlights searched the sky. The observer aboard W-45 recorded at 0010 that they had been caught in beams of four searchlights, and five minutes later they escaped them by heading out to sea.

Both Emilys headed for home at 0040 after dropping a total of fifteen x 250-kilogram bombs on the wharf where three merchant ships were berthed. However, the bombs all detonated harmlessly in the water. The sloop HMAS *Swan* was anchored in the harbour with one boiler stripped for cleaning. After receiving the air raid warning at 0001, the sloop was able to get underway on the other boiler twenty minutes later. *Swan* then had a minor collision with the old coal-burning cargo steamship *Time* as both vessels hastened to vacate the harbour.

Signals were sent to No. 76 Squadron at Milne Bay, so that its P-40E Kittyhawks might intercept the Emilys on their return to Rabaul, but take-off was delayed by an hour for unspecified reasons and no interception took place. The long fifteen-hour mission ended for the Emilys when they alighted back at Simpson Harbour at 0712 that morning.

After a day's respite, on 27 July 1942 Lieutenant Mizukura Kiyoshi and co-pilot Takeya Takeo left Rabaul in W-46 shortly after dusk. They arrived over Townsville at 0225 early the next morning, however this time RAAF radar had plotted their approach about two hours out. Six 8th FG Airacobras launched from Garbutt airfield and were airborne by the time the flying boat was approaching Townsville, however they failed to locate it. Searchlights assisted AA batteries in targeting the Emily, and 72 rounds were fired but without result. Mizukura dropped his eight 250-kilogram bombs from 15,000 feet over Garbutt. These landed near the airfield but did no damage. Mizukura was safely back at Rabaul later that morning after another fifteen-hour mission.

The next day of 28 July 42 saw the third and most eventful Townsville raid, when W-37 and W-47 departed Rabaul. However, W-37 soon returned to base with unspecified technical problems, leaving W-47 under the command of Lieutenant Shoji Kingo to continue alone. After reaching Townsville at 0025, the crew was greeted by ten searchlights. On this occasion the AA batteries had agreed to remain silent to enable fighter interception.

This time four 8th FG P-400 Airacobras had received thirty minutes radar warning and were waiting at altitude when the Emily approached. Aided by the searchlights and a clear evening lit by a full moon, Captains Robert Harriger flying BW 183 and John Mainwaring in BW 163 attacked the flying boat. Shoji dived to 18,000 feet in an attempt to escape, later logging that at 0028 they were attacked by two "Hurricanes" in seven separate passes during a running battle of some fifteen minutes, and noting his gunners had hit one of the attackers more than ten times.

Harriger pursued the Emily past the searchlights but soon ran low on fuel and ammunition forcing his return to Garbutt. Harriger thought that on his last pass he had started a small fire in the tail of the Emily which burned only briefly. During the pursuit Shoji dumped seven 250-kilogram bombs into the harbour between the shore and Magnetic Island. In response, HMAS Swan opened fire with her 4-inch anti-aircraft guns.

Meanwhile an eighth bomb exploded near Townsville racecourse and shattered several nearby windows. The next morning gatherings of curious locals inspected the resultant large crater. Mainwaring was later quoted in the local Townsville newspaper:

> If we had been a bit less excited I guess we could have made sure of it. We came in together and got our first burst home from the tail, along to the underpart of the hull. Bob was a little cramped for space, so he slipped underneath me, and I could see his tracers coming up and hitting the side. I got in so close that I was afraid I would collide with her. She was a big ship, and, stuck out there in a pool of lights, she looked like a model. There was not a shadow on her anywhere. Our first burst silenced the rear gunner, and one of our shells seemed to explode inside the ship. Afterwards only the top turret kept firing.

In fact, Shoji's Emily had received only slight damage and arrived safely back at Rabaul shortly before dawn. After this third raid Radio Tokyo broadcast jubilantly, but not without hyperbole that:

> All-important military installations at Townsville were smashed in three raids by Japanese naval air units. On 25 July airfields, oil tanks, shipping and supply dumps were raided. On 28 July airfields, oil tanks and supply dumps were attacked and on 29 July the remaining military installations were bombed. This attack on Townsville was one of the heaviest since the fall of Singapore.

The efforts of No. 14 Ku against northern Queensland continued against two other locations during three consecutive nights. The first of these raids was against Horn Island by a single

Emily. At 0210 on 30 July seven 250-kilogram bombs were dropped accurately over the airfield dispersal area where a No. 32 Squadron Hudson was damaged with three others lightly damaged.

The following night another Emily targeted Cairns, but in fact mistakenly bombed the small town of Mossman, some 40 miles to the north. One bomb damaged the home of a sugar cane grower, slightly injuring his infant daughter, who became the sole casualty from this series of raids. A second raid against Horn Island was made on the night of 31 July / 1 August by a pair of Emilys. An air raid alert was raised at 0315 but no raid was subsequently recorded. In fact, only one Emily released its bombs all of which fell harmlessly into the sea.

While occasional Japanese reconnaissance flights over northern Queensland continued, only one further raid took place over that state almost a year later, in June 1943. Nonetheless the alleged success of the Emily raids against mainland Australia had strategic consequences unknown to the Allies. Back in the corridors of Japan's military headquarters, much was made of the Townsville raids by the IJN, highlighting their exclusive ability to wield a long-range strategic bombing capability. This fuelled further ongoing IJA / IJN rivalry, which in turn hastened JAAF plans to demonstrate their own ability to hit Australia. To do this in mid-1943 they deployed Ki-49 heavy bombers to Timor from where they raided Darwin in June 1943. Then, to again underline JAAF long-range offensive capacity, on the evening of 20 September 1943 four Ki-49s launched from Wewak to bomb Port Moresby, of which only two got through. This marked the final Japanese air-raid against the Papuan capital for the entire war.

A wartime view of Townsville's harbour and jetty which was targeted by Emily flying boats during a series of night raids in late July.

Two Australian soldiers search for Japanese bomb fragments following the third raid on Townsville. Despite extravagant Japanese claims, no serious damage was inflicted by the raids.

From Lae on 25 July, No. 4 Ku officer Lieutenant Obata Toshikatsu organised a series of low and medium-level strikes against Mubo and Gabmatzung Mission, marking one of the few times the G4M1 Betty was used as a strafer.

CHAPTER 7

150 BOMBERS DESTROYED! NEW GUINEA 1–8 AUGUST

The new month was only a few hours old when Japanese aircraft bombed Seven-Mile between 0230 and 0300 on 1 August. Cloud prevented observation of the type involved although AA still managed to fire 224 rounds. Some bombs landed within the northwest corner of Seven-Mile, but no damage was done. This was another No. 4 *Ku* nuisance raid that went awry. Two *shotai*, each of three Bettys, was intended to strike Port Moresby about an hour apart. However, after encountering bad weather about two hours into their mission the first *shotai* returned to base. The second *shotai* reached the target area and as noted above each Betty released 12 x 60-kilogram bombs over Seven-Mile in staggered intervals. The third Betty in this second *shotai*, flown and commanded by FPO1c Sekine Tokushiro, never returned and remains missing, almost certainly a victim of bad weather over New Britain.

Such were the emergencies soon to develop elsewhere in the theatre, it would be more than two weeks before Japanese aircraft appeared over Port Moresby. This was the largest rest period the town had faced since raids began in late February. Indeed, since that time just five months earlier, Port Moresby had been raided 76 times. While on only a few occasions did these raids inflict significant damage and losses on the Allies, they had a major deterrent effect in preventing the permanent basing of medium or heavy bombers at Port Moresby. This in turn seriously diluted the effectiveness of Allied airpower, due to many factors, and surely counts as a significant win for the Japanese in the 1942 New Guinea air campaign.

Meanwhile, efforts to reinforce the beachhead area were continuing. At 0700 on 31 July three ships comprising the fourth Buna transport convoy departed Rabaul as shown:

Fourth Buna Convoy

- *Tsugaru* (4,400-ton mine-layer; built 1941; 4 x 5-inch guns; 4 x 25mm guns; 1 x E7K Alf floatplane; used as AA ship)
- *CH-28* (420-ton submarine chaser; built 1942; 1 x 3-inch gun)
- *Nankai Maru* (5,114-ton IJN auxiliary transport; built 1933)

This convoy was spotted by Allied reconnaissance late that morning off the south coast of New Britain and was somewhat accurately surmised as "one medium cargo vessel, one warship and one launch". The following day, 1 August, was another day of weather sufficiently problematic to ground all Lae-based Tainan *Ku* Zeros. A flight of B-17s attempted to bomb the ships, with one reporting missing a "10,000-ton cargo vessel" from high altitude. Nevertheless, this attack was enough to convince the convoy commander that the run to Buna was too risky in light of recent losses, and that night the ships turned around.

On 2 August the weather had cleared sufficiently for the Sasai *chutai* of nine Zeros to depart from Lae to fly more protective patrols over Buna. These Zeros encountered five 28th BS Fortresses flying at low altitude underneath the overcast that were searching for the ships. Sasai's pilots made aggressive frontal attacks, and at 0930 they were successful in shooting down the B-17E flown by Lieutenant William Watson near Cape Ward Hunt. Several crew bailed out, but only Sergeant Leo Ranta survived - he was subsequently rescued by Australian soldiers. After many attempts, the Tainan *Ku* had finally removed a Fortress from the skies of New Guinea. Sasai himself claimed the bomber, although the Tainan *Ku* themselves lost the newly arrived Flyer1c Motoyoshi Yoshio to Fortress guns during this engagement.

Approaching the area at this time were five 22nd BG B-26s that were escorted by a dozen P-400s from the 41st FS. However, both the Marauders and Airacobras had got separated in the weather, with the result that a trio of Airacobras arrived first (the fate of some others is described below). The 41st FS was still a relatively inexperienced unit, and the Americans came off second best when set upon by Sasai's elite *chutai*. After a brief combat, two Airacobras were downed. Lieutenant Jesse Dore likely crashed into the sea, but the fate of Lieutenant Jesse Hague remains obscure. After bailing out Hague joined a mixed group of Allied fugitives inland from the Buna area and was last seen exchanging fire with Japanese troops, although he was not among the captives executed on Buna Beach on 12 August. He remains missing to this day.

The Zeros now turned their attention to the five unprotected B-26s, causing the Marauders to scatter. One of the B-26s, named *Our Gal*, jettisoned its bombs and became the focus of a *shotai* of two or three Zeros. After damaging the B-26 in the right wing, the Zeros broke off and returned to Lae. *Our Gal* was probably saved by the Zeros running short of ammunition and possibly fuel. On return to base the eight surviving Zeros tallied an expenditure of nearly 6,000 rounds of ammunition, although six of the fighters had each sustained bullet hits in return.

Later in the day another patrol of five Zeros led by Lieutenant Yamashita Joji sparred with a lone 435th BS B-17E near Buna. These Zeros also expended a considerable amount of ammunition (3,900 rounds), but there were no losses to either side. Meanwhile the *Tsugaru* had reported engaging a single B-17 in the morning and another in the afternoon, firing both its 5-inch and 25mm guns. All three ships of the convoy had returned to Rabaul safely by 3 August.

In a day of firsts, it remains to describe an encounter several of the 41st FS Airacobras had early that morning when they became separated from the Marauders they were escorting. These had chased a Tainan *Ku* J1N1-C Irving twin-engine reconnaissance aircraft which had recently entered service with the unit. It had departed Lae for a morning mission over Port Moresby, flown by Warrant Officer Tokunaga Tamotsu. When the unsuspecting Irving was downed by Lieutenant Elbert Schinz, it became the first example of the type to be lost in combat anywhere.

Meanwhile Major-General Kenney had arrived in Australia on 28 July, having learned of the Buna landings when passing through Hawaii on his trans-Pacific flight. After meeting MacArthur, Kenney was informed of the Tulagi and Guadalcanal invasions planned for 7 August and the need for the SWPA bomber force to give maximum assistance. Immediately Kenney borrowed Brett's personal transport plane, an old B-17D named *The Swoose*, and flew to Port Moresby

The navigator of a G4M1 Betty checks his position on the way to Port Moresby, which was raided 76 times between February and 1 August 1942.

and then to Mareeba to inspect the 19th BG. Kenney formed a low opinion of the unit, writing in his memoirs that:

> The crews were thinking only of going home ... Their morale was at a low ebb and they didn't care who knew it.

Kenney was appalled to learn that of 32 B-17s on strength, only fourteen were ready for operations and most were "pretty much worn out". Kenney famously temporarily stood down the tired group so as to have as many B-17s as possible available to use against Rabaul to coincide with the Solomons landings planned for 7 August. At this time Brett was sent back to the US aboard *The Swoose* and Kenney assumed command of the Allied Air Forces, SWPA. Criticism of Brett and the 19th BG by both MacArthur and Kenney ironically coincided with when the mere appearance of B-17s had forced the fourth Buna convoy to turn back in a small but significant victory for the Allied Air Forces.

During Brett's return flight across the Pacific to Washington DC *The Swoose* set multiple speed records. The aircraft was used by Brett as his personal transport until late 1945, after which it narrowly escaped scrapping and was subsequently passed between various museums and display locations. Today *The Swoose* is the oldest surviving B-17 and is undergoing restoration at the National Museum of the United States Air Force.

After months of doggedly defending Lae, the Tainan *Ku* detachment there received orders to withdraw to Rabaul on 3 August. The relief was palpable, as Lae suffered the threat of surprise Allied air attack at any time. To combat this the Tainan *Ku* pilots maintained lengthy Combat Air Patrols at every opportunity, but these were draining both for the machines and their pilots. Without the benefit of radar or a deep network of physical spotting stations, the Zero was far less efficient in defence than attack. Further, the IJA believed the IJN had failed in its duty to provide air cover over the Buna beachhead area – weather and other considerations notwithstanding. However, the Buna airfield was almost ready for operations and would soon house its own detachment of Zeros.

One of the reasons for the withdrawal from Lae was a need to consolidate Tainan *Ku* strength for another offensive operation planned against Samarai on the far eastern tip of Papua. For this purpose, a reconnaissance was ordered on 3 August, which was flown by a No. 4 *Ku* Betty. This mission discovered the Allied airbase at Milne Bay for the first time, which the Japanese named Rabi after a village in the same vicinity. Ironically, the Milne Bay airfield site was just a short distance from Samarai – the Allies had beaten the Japanese to it. The Betty crew noted a well-developed runway, and accurately counted the presence of around 40 Kittyhawks. Jetties had been constructed in the bay with a medium sized transport alongside. A detailed chronicle of activities at Milne Bay from this juncture follows in Chapter 10.

After the loss of Kokoda and its airstrip, some 550 Australian troops, mostly from the 39th Battalion, were located forward on the mountain trail which would soon be known simply as the Kokoda Track. The problems of supply were daily becoming more acute. While many hundreds of native carriers were being organised, it was realised that air supply was the only means by which reasonable reserves of food and ammunition could be accumulated in the mountains. To this end the Australian Army blazed a new trail to some grassed dry lake beds high in the mountains that were named Myola.

Soon regular supply drops were being made to this location, by both the handful of transport aircraft available and No. 32 Squadron Hudsons. With little specialised expertise available, many of the initial drops burst on impact with supplies lost or destroyed. However soon supplies were packed into multiple inner and outer layers of sacks. The inner layers would burst on impact but usually the more loosely packed outer layers would contain the contents. Nevertheless, the type of supplies that could be dropped was limited to items which could survive the impact.

Along with the supply dropping missions, the fighters at Port Moresby were also being used regularly in direct support of the land forces. Four Airacobras bombed and strafed Kokoda at 0930 on 2 August, reporting hits on the former District Officer's house. Despite bad weather the following day, several Airacobras were again in action and strafed and destroyed grass huts at Kokoda and Oivi, where Japanese supply dumps were suspected.

Taking advantage of the relative lull in operations on 4 August, Airacobras were engaging in dive-bombing practice against the wreck of the SS *Pruth* outside Port Moresby harbour that had foundered on the reef in 1923. During one of these practice runs the CO of the 39th FS, Major Jack Berry, was killed when his fighter dived into the ocean. That morning Berry was carrying a 250-pound practice bomb and was supposed to hit the shipwreck. Bombing with a P-39 was a new concept at the time. A gun crew on Paga Hill above Port Moresby saw something unidentified come off the plane just before it disappeared from sight.

On this same day, 4 August, an air reconnaissance of Rabaul reportedly showed 150 aircraft massed at the airfields there. As will be explained, the Japanese had recently received bomber reinforcements, although the supposed estimate of 150 was a considerable exaggeration. Nevertheless, with much tension among Allied leadership about the upcoming Solomons landings just three days away, the high bomber count led to speculation as to whether the Japanese had somehow learned of the operation.

Since the start of the month RAAF Catalinas had been flying night nuisance raids over enemy locations. The first of these was by A24-22 and A24-26 in the early hours of 2 August having departed Cairns late the previous afternoon. After finding the Gona area completely covered by

A crewmember of B-17D The Swoose points to the fuselage motif while the bomber is being serviced at Laverton, Victoria. The Swoose flew Lieutenant-General George Brett back to the US in early August after he was replaced as Allied Air Forces commander by Major-General Kenney.

an overcast with a cloud base of only 500 feet, bombs were dropped over Lae from 600 feet but with no observed results. The following night saw slightly better conditions which allowed A24-10 and A24-12 to bomb Lae and Salamaua from 4,800 feet, but again without observed results.

Just prior to the Solomons "D-Day", attacks were launched against Lae and Salamaua with the intention to keep Zeros away from Port Moresby where the B-17s were assembling. Seven B-26s attacked Lae on the morning of 6 August, with the bombs falling on the runway: men and trucks were later seen repairing damage. At the same time six B-25s bombed Salamaua's small grass airstrip. The Allies had yet to learn that Lae had been vacated by its Zero detachment and the small field at Salamaua was only used as an emergency alternate to Lae. These attacks were repeated on D-Day itself, with ten tons of bombs unloaded over Lae in the biggest attack on that target yet. Overall, the attacks continued for three days until 8 August and involved 56 sorties in total: 38 by Marauders, 13 by Mitchells, three by Catalinas and two by Fortresses.

The sorties above included missions on the night of 6-7 August by three Catalinas which bombed Lae. These were the first in a busy series of night sorties by the RAAF flying boat force. The following night Catalina missions were flown against Lae (by one aircraft), Rabaul (two) and Buka (two). The missions against Rabaul were labelled as a nuisance raids intended to prevent night repairs to the airfields at Vunakanau and Lakunai following the daylight B-17 raid. One of the Buka attacks was flown by Flight Lieutenant GR Thurstun in A24-4 who encountered very adverse weather but was able to light up the runway area after dropping flares from just 1,500

A Marauder has its engine changed at Antil Plains, Queensland, where an enormous amount of maintenance was performed to keep the B-26 force flying.

feet. No enemy aircraft were observed and after bombing the runway the Catalina received a single bullet hole in its tail before returning to Cairns at lunchtime on 8 August.

However, the all-important mission was the maximum-effort B-17 raid against Rabaul on 7 August. Late on 6 August a total of fifteen B-17s had concentrated at Seven-Mile, five each from the 28th, 30th and 93rd Bombardment Squadrons. In addition, the B-17E flown by Captain Harl Pease of the 93rd BS had just blown a cylinder valve. Not wanting to be left out of this landmark mission, Pease flew back to Mareeba to search for another serviceable Fortress. He found an older "E" model named *Why Don't We Do This More Often?* that was suffering from worn engines and recurring electrical problems. Pease was determined, however, and flew the bomber to Seven-Mile that night, arriving at 0100 on 7 August.

At first light the following morning, the 19th BG commanding officer, Colonel Richard Carmichael, from the co-pilots seat of the unnamed 41-9015 led the sixteen Fortresses one-by-one down Seven-Mile's runway. However, a runaway supercharger led one bomber off course during the take-off, and it crashed into a rock pile and was destroyed. Two others soon turned back due to technical problems, leaving Carmichael to lead thirteen bombers towards the target area. As they approached Rabaul at 1130 the Fortresses were bathed in bright sunlight.

Little did the Americans know that the best of the Tainan *Ku*, some eighteen pilots, were at that time away on a mission to Guadalcanal. However, another unit, the No. 2 *Ku*, had just arrived at Rabaul (as described below) and managed to scramble fifteen Zeros led by their *hikotaicho* Lieutenant Kurakane Yoshio at 1115. These were joined five minutes later by eleven Tainan *Ku* machines.

As the Americans dropped fifteen tons of bombs over Vunakanau, they were swarmed by

Newly appointed commander of the Allied Air Forces, General George Kenney (left), with his bomber commander Lieutenant-General Kenneth Walker at Port Moresby. Kenney's claim that between 75 and 150 enemy bombers were destroyed during the 7 August B-17 raid was one of the most extraordinary over-claims of the entire Pacific War.

the 26 Zeros. Among those targeted was Carmichael's lead Fortress, where one of the gunners was killed and the oxygen system was ruptured. Carmichael immediately radioed to the others that he had to descend, and his two wingmen followed him. However, this action by the lead trio broke up the mutual defensive firepower of the Fortresses and left stragglers vulnerable to be picked off.

In this dangerous situation was Pease with his underperforming bomber, which was last seen with one engine windmilling and surrounded by Zeros. An attempt was made to jettison the blazing bomb bay tank, which fell loose but the flames had already engulfed the stricken Fortress. This was the only loss during the mission although other Fortresses sustained damage. The No. 2 *Ku* correctly claimed one B-17 shot down after expending 4,276 rounds of ordnance, although Tainan *Ku* pilots likely also contributed to the kill. Pease and one of his gunners, Sergeant Chester Czechowski, managed to parachute safely and were the only survivors from *Why Don't We Do This More Often?* Both were captured and subsequently executed at Rabaul.

Kenney and MacArthur believed the raid had been highly successful and contributed significantly to the success of the Guadalcanal landings that same day. However, the truth is most of the Vunakanau-based Bettys were in the air at the time attacking Allied shipping off Guadalcanal. Not a single Japanese aircraft was destroyed although repairable damage might well have been inflicted to Vunakanau's runways and taxiways. In one of the most outlandish embellishments of the entire Pacific War, Kenney claimed that between 75 and 150 enemy bombers "parked wingtip to wingtip" were destroyed during the raid. It is difficult to know how Kenney gleaned this assessment, for reconnaissance photos could not have showed any inkling of damage. Furthermore, in post-war interviews of Japanese commanders at Rabaul at the time, the 7 August raid barely features in their collective memories.

Despite this overclaim, Kenney nonetheless understood the value of morale and confidence to his young flyers. He subsequently instilled the belief that they had struck a major blow against the enemy leading to a considerable morale boost to the tired 19th BG aircrews. Kenney was also liberal with the handout of awards, including a posthumous Medal of Honor awarded to Pease, presented to Pease's father by President Roosevelt four months later. The award remains the sole Medal of Honor won by the 19th BG.

A6M3 Model 32 Zero, CN 3035, Q-101, HK-877, No. 2 Ku, Hikotaicho Lieutenant Kurakane Yoshio, August 1942.

While the Rabaul B-17 mission was underway, an epic story of survival was just beginning in the jungles of Papua. That afternoon, 7 August, a flight of three Marauders had taken off from Woodstock in Queensland bound for Port Moresby. Each Marauder carried a 1,000-pound bomb intended for use against Lae. The trio was led by Lieutenant Robert Hatch in *DIXIE* with wingmen Lieutenant Seffern in *Yankee Clipper* and Lieutenant Albert Stanwood. Approaching New Guinea the Marauders ran into solid cloud, with Hatch trying to find a way over it. Stanwood recalled:

> we ran into bad weather, and after it became obvious to me that Hatch was lost, I left the formation, headed south to take me out to sea, descended through the cloud, broke clear over water at five hundred feet, then turned north. I was able to find Moresby by knowing to turn left if I came to the reef, or right if there was none. I pulled up at the shoreline, spotted the runway and landed just as night was falling.

Meanwhile, with dusk falling the other two crews were in serious trouble. Seffern knew that he was flying towards an area dotted with mountain peaks towering to 14,000-feet, the minimum safe altitude, and he ordered his men to bail out. This was complicated by the fact that a civilian war correspondent was aboard, Verne Hughland. After bailing out from such a height the men were scattered over a wide area. Over a period of days most staggered into the same village from where they were led to the coast and made contact with Australian soldiers. Just two men, the co-pilot and navigator, were never seen again. At first there was no trace of Hughland either, but most surprisingly he reappeared in Port Moresby months later, suffering from exposure and close to exhaustion. Hughland later wrote a book about his experience.

While Seffern had elected to bail out, Hatch had the philosophy of keeping his crew together and decided to crash-land. After jettisoning the bomb, he found a hole in the cloud and descended. He then realised with some dismay that he was over the Japanese-occupied north coast of Papua. Hatch then made a successful belly landing in a swamp, not far from the coast and some miles to the south of Buna. Remarkably, the crew spent weeks in an area evading Japanese patrols before meeting friendly forces and being flown out of a mountain strip by a RAAF Hudson. Hatch's intact bomber today lies buried in mud.

As noted above, the B-25 and B-26 missions against Lae and Salamaua continued until

8 August without further incident. Among other routine operations, sixteen Airacobras undertook strafing attacks near Kokoda. Allied intelligence optimistically reported that these attacks silenced machine gun posts and set huts on fire. However, without direct coordination with friendly ground forces it was very difficult for fighter pilots to target the enemy under the jungle canopy. In fact, such attacks accomplished very little throughout the entire Kokoda campaign, substantiated by the copious amount of Japanese primary-source documentation which surprisingly survives on the topic.

On 3 August, an RAAF Catalina dropped supplies to coastwatcher Jack Read and several men of the 1ˢᵗ Independent Company still hiding out in northern Bougainville. At this time Read reported that enemy flying boats were beginning to use Soraken as an overnight anchorage and base from which to patrol the Bougainville coast. Named "Saipeso" in *katakana* by the Japanese, the anchorage was an abandoned RAAF AOB southwest of Buka Passage. Lieutenant Narita Yoneo was sent from the Yokohama *Ku* base at Malaguna, Rabaul, to first survey the anchorage on 5 August, putting down at Saipeso at 1420 where he stayed overnight and returned to Rabaul the following day. Also noted by Read was construction activity at Buka airfield, where hundreds of natives had been conscripted for the task. A few isolated landings had been noticed on the airstrip.

These reports led to a number of subsequent Allied reconnaissances of the area and occasional bombing missions. On the afternoon of 7 August Catalina A24-25 of No. 11 Squadron departed Cairns at 1500 for a night attack against Buka airfield.

Floatplanes from *Kiyokawa Maru*, although predominantly based in the Papuan area to support ongoing Buna operations nonetheless were broadening their presence throughout the region in early August 1942. Aside from those still aboard the ship, three additional Aichi Jakes were based at Kavieng (having transferred from Truk, but now attached to the seaplane tender) and four more Jakes were at Malaguna Bay, Rabaul. Furthermore, the ship's floatplanes also sometimes operated out of the Shortland Islands, Tulagi and Gizo in the Solomons.

Meanwhile the Japanese had been receiving some welcome reinforcements. On 29 July the 11,300-ton seaplane tender *Nisshin* arrived in Rabaul. Built to carry 24 floatplanes, the vessel had a load of twenty Model 32 Zeros strapped down as deck cargo instead, these being the first Model 32s to arrive in the theatre. With distinctive squared-off wingtips, the Model 32s offered a marginal increase in power compared to the existing Model 21s but had a much-reduced range. For this reason, they would largely be tasked with defensive duties at Rabaul and later at Buna with the Tainan and No. 2 *Ku*. While still on deck the Zeros were noted by Allied reconnaissance on 29 July as being "21 planes on the flight deck of an aircraft carrier". The search for this supposed carrier briefly occupied the minds of Allied planners before resources were concentrated against the fourth Buna convoy which was spotted on 31 July as described at the start of this chapter.

A week later a real aircraft carrier arrived in Rabaul. This was the newly completed escort carrier *Unyo* which was being used as an aircraft transport. The *Unyo* delivered a consignment of Zero fighters and Val dive-bombers belonging to the No. 2 *Kokutai*, which was a mixed fighter / dive-bomber unit. The No. 2 *Ku* had been newly formed in Japan on 31 May with an establishment strength of sixteen each of Zeros and Vals under the overall leadership of Commander Yamamoto Sakae. The unit was intended to be based at Port Moresby after the capture of that location.

This first batch of No. 2 *Ku* Zeros comprised fifteen Model 32s, each of which were *hokoku* aircraft that had been donated by patriotic organisations or individuals. Their unique *hokoku* markings had a multicultural factor too, with all of the designations from HK-870 through to HK-878 (and possibly others) being Korean donations. This aspect still exercises confusion among historians as to their translation due to the Korean *kanji* characters used.

The No. 2 *Ku*'s Vals were sixteen early model D3A1s which had already seen service in other units. However, these land-based D3A1s, unlike their carrier-based contemporaries, lacked auxiliary fuel tanks. Neither did they have the same under-fuselage hardpoints upon which the carrier-based Vals carried 250-kilogram bombs. Instead, No. 2 Ku's Vals were limited to carrying a pair of under-wing 60-kilogram bombs which offered only modest destructive power.

Subsequently No. 2 *Ku* would be deployed to Buna, but as described above its fighters first saw action on 7 August over the skies of Rabaul.

Further reinforcements were flown into Vunakanau from the Central Pacific on 8-9 August, comprising 27 Bettys of the Misawa *Kokutai*. The unit had been formed in Japan in February before moving to Saipan in July. The Misawa *Ku* was part of the 26th Air Flotilla which had been earmarked for service at the newly completed Guadalcanal airfield. In fact, the first Misawa *Ku* bomber had arrived at Vunakanau on 23 July, this being a solitary G4M1 which delivered staff officers to set up their new unit headquarters. Another solitary "airlift" bomber followed two days later.

As mentioned in Chapter Two, the small coastal submarines *RO-33* and *RO-34* were attached to the Eighth Fleet and had been allocated for operations south of Port Moresby. Both vessels departed Rabaul on 29 July. The *RO-34* briefly operated off the east coast of Cape York before being redirected to the Solomons after the American landings on 7 August. The *RO-33* meanwhile was operating in the Gulf of Papua where at 1034 on the morning of 7 August she came upon the 300-ton Burns Philp motor vessel *Mamutu* on a voyage from Port Moresby to Daru.

Over 100 people were aboard the *Mamutu*, including a mixed European / native crew and many native passengers who were being resettled in Daru. When the submarine began surfacing nearby, the *Mamutu* attempted to escape and the radio operator was able to broadcast a submarine warning message. However, the *RO-33* quickly caught up and from a range of 300 yards crewmen had unlimbered the 3-inch deck gun and began firing. The first shells wrecked the radio room and bridge and killed the captain, Master J McEachern.

As further shells hit home within minutes the *Mamutu* was a blazing wreck, and many survivors were in the water clinging to debris or aboard a partially submerged lifeboat. The *RO-33* then

The Australian coastal vessel Mamutu which was sunk with heavy loss of life in the Gulf of Papua by the submarine RO-33 on 7 August.

circled the area with its crew machine gunning survivors in the water before submerging, leaving just a few survivors in the damaged lifeboat.

On receipt of the distress message, various aircraft were alerted to search the area, including No. 32 Squadron Hudsons A16-186 and A16-205 which departed Port Moresby that afternoon and after searching for four hours landed at Horn Island that evening. The only aircraft to sight the survivors was Empire Flying Boat A18-11 of No. 33 Squadron and pilot Flight Lieutenant Michael Mather decided to land despite rough seas. However, the flying boat was damaged by the large waves and sank quickly, with the pilot and seven crewmen taking to a rubber lifeboat. An eighth crewman, Leading Aircraftsman George Edwards, was killed.

Subsequently one native survivor swam from the damaged *Mamutu* lifeboat to the rubber lifeboat containing the downed RAAF crew. After drifting for a few days, the rubber lifeboat subsequently reached land, but the other lifeboat was never seen again and this one native survivor on the rubber lifeboat became the sole surviving witness to the loss of *Mamutu*. That afternoon *RO-33* was redirected to the Solomons along with *RO-34*.

The following day another submarine action unfolded in waters off Rabaul. This was in relation to a small convoy despatched from the port on 7 August carrying 519 SNLF marines. These had been intended to reinforce the newly built airfield on Guadalcanal but were now being rushed to contend the American invasion that morning. The marines were loaded on the 5,627-ton naval transport *Meiyo Maru* and the ammunition ship *Soya*. Escort was provided by the minelayer *Tsugaru* and two smaller vessels including the minesweeper *W-21*.

However the following day, after learning about the size of the US invasion force, Vice-Admiral Mikawa recalled the convoy. As the ships approached Rabaul shortly before midnight that evening, they were spotted by the American submarine *S-38*. Two torpedoes hit the *Meiyo Maru* which went dead in the water and began sinking within minutes, with the loss of 342 marines and 31 crewmen. The *S-38* was depth charged but escaped unharmed. The submarine remained in the area until 15 August but didn't encounter any further targets and returned to Brisbane.

A Rufe floatplane from the Yokohama Ku's fighter detachment attacks an 11th BG Flying Fortress over Tulagi in early August 1942.

CHAPTER 8

D-DAY
THE SOLOMONS 1–8 AUGUST

The first week of August saw "D-Day" for the landings at Tulagi and Guadalcanal fast approaching on 7 August. Accordingly, a keen watch was kept on the area with regular reconnaissance flights. In addition, the newly arrived 11th BG was ordered to bomb the Guadalcanal airfield as much as possible in the hope that its completion would be delayed at least until the landings took place.

Following the first mission by the 11th BG on 31 July, again using the forward bases in the New Hebrides, the group sent ten Fortresses against Guadalcanal and the Tulagi seaplane mooring area on 1 August. At 1000 the first patrol of six Rufes was airborne from Tulagi led by *chutaicho* Lieutenant Sato Ri'ichiro. Two separate skirmishes occurred with the Fortresses at 1020 and 1130, but with no losses to either side. Two moored Mavises received damage that was judged to be repairable.

However the rudimentary conditions at the forward airfields in the New Hebrides claimed their first victims at 0100 on 2 August when a B-17E named *Jap Happy* clipped a tree on take-off and then collided with a parked B-17E of the 26th BS (41-9155). The stationary bomber was wrecked while *Jap Happy* would be non-operational for some time. A few hours later two more taxying 26th BS B-17s brushed wingtips, putting both out of action and ending plans for a co-ordinated morning strike.

Instead two B-17s which succeeded in taking off in the darkness proceeded alone but soon returned to Efate in the face of gloomy weather. Just before dawn another four 26th BS B-17s took off and these bombed Guadalcanal successfully with 21 x 500-pound and 14 x 300-pound bombs, claiming:

> Hits were scored with 75% of the bombs – runways were smashed and two hangars left burning.

Following shortly afterwards was a pair of 98th BS B-17s which were intercepted by a dozen Rufes at 0920, causing serious damage to the

Nose art on 11th BG B-17E Jap Happy which clipped a tree on 2 August while departing Efate during the early raids against Guadalcanal.

right outboard engine of one of the bombers which left the area trailing smoke. Two Fortresses were claimed as definites by the Yokohama *Ku*, in exchange for light damage to one of their Rufes. This was a substantive battle too, with the Rufes expending 1,150 x 20mm and 2,200 x 7.7mm rounds of ordnance.

On 3 August, only a pair of 26th BS B-17s could be mustered to strike Tulagi from 6,000 feet, but no damage was recorded by the Japanese. A trio of 431st BS Fortresses from Fiji was also ordered to attack the airfield, via Espirito Santo. It is possible they didn't reach the target due to the weather; unfortunately the detailed records for the Yokohama *Ku* Rufe detachment cannot be cross-checked as the records from this day onwards were destroyed during the forthcoming US invasion (more generalised Japanese records at fleet level are available throughout the invasion which reference Yokokama *Ku* activities).

The 11th BG made a bigger effort against Guadalcanal on 4 August, when three trios from the 26th, 98th and 431st Bombardment Squadrons made separate attacks. The 26th BS trio was first to arrive over the target area but much to the disgust of the crews the trailing trio from the 98th BS were not maintaining radio silence. Instead, idle chatter between crews betrayed their approach to the Solomons.

According to the 26th BS crews this forewarned the Japanese such that six Rufes were waiting for them which dived on the three Fortresses. A floatplane flown by Sea1c Kobayashi Shigeto was hit by defensive fire from the bombers and caught fire before crashing into unnamed B-17E 41-9218 flown by Lieutenant Rush McDonald. This was witnessed by Lieutenant Jack Lee who wrote:

> ... seven float-type Zeros waited for us. They worked around for a frontal attack and made their first try before our bomb run. Two of the first group dived and cut loose with their guns in turn. The third was killed in his dive by the top turret gunner. He was in a vertical bank and his right wing hit McDonald's ship between the two LH engines. The Zero exploded and blew the wing of the B-17. I turned in my seat to watch the Zero as it went by and saw the crash. There was terrific flame from the burning gasoline. The 17 turned on its side with the stub up then nosed down and fell out of sight. The Zeros continued to make passes as we dropped our bombs and turned down the coast of Guadalcanal towards home. A very tough day.

An IJN launch raced to the crash site from Tulagi and found Kobayashi's helmet and clothing in the water. The nine crew aboard the B-17 were all lost, however two American airmen were captured by the 14th Establishment Unit on Guadalcanal that day. Doubtless from McDonald's crew, they were likely quickly interrogated and killed.

On 5 August the effort continued with three 26th BS and seven 431st BS B-17Es bombing Tulagi, their targets including a hospital which the Americans believed was being used to store munitions. Japanese sources state that nine Rufes doggedly attacked the bombers, although there were no losses to either side.

More air strikes were scheduled on 6 August, but a driving rainstorm and trouble with the only mechanised fuel pump at Espirito Santo kept the planes grounded. All available hands, including the 11th BG commander Colonel Saunders and the ground forces commander,

Boeing B-17E, 41-9218, 26ᵗʰ BS, 11ᵗʰ BG, rammed by a Rufe floatplane on 4 August 1942 and destroyed.

Brigadier General William Rose, worked a bucket line which reportedly transferred 25,000 gallons of fuel. However, that was still not enough for a large strike formation, and instead various B-17s flew missions to the north of Guadalcanal searching for any Japanese naval presence that might interfere with the forthcoming landings, but nothing was spotted.

A handful of solo bombing missions were managed on 6 August, and on one of these missions Captain Rolle Stone flying 41-9221 returned in darkness. Unable to find Espirito Santo's airfield, he ditched his 42ⁿᵈ BS Fortress offshore from the northern end of the large island. Fortunately, all nine crewmen were uninjured but another Fortress had been lost.

On D-Day itself, 7 August, the 11ᵗʰ BG was held at readiness, but with the Guadalcanal landing being unopposed the Fortresses were not required. A few 98ᵗʰ and 431ˢᵗ BS machines were required for search missions to cover the northern Solomons and further afield towards Rabaul. However, a pair of Tulagi-based Mavis airborne on their own extended sector patrols from 0600 was the only Japanese activity in the region and nothing was spotted by the B-17s.

One of the Fortress missions was flown by Major Marion Pharr, the CO of the 431ˢᵗ BS, who departed Espirito Santo at 0300 on 7 August in B-17E 41-2426 but never returned. A tail gunner in another Fortress heard one of Pharr's gunners broadcast "We're hit. We're going in", indicating a strong possibility that Pharr was shot down by Japanese fighters. The only such fighters in the area were the Yokohama *Ku* Rufes, however the unit was wiped out on this day, as explained below, and all records were lost. Pharr's loss is sometimes listed as possibly due to friendly fire as this was the day of the invasion and much confused activity took place.

Further south, the 11ᵗʰ BG lost another Fortress on this date, when the CO of the 98ᵗʰ BS, Major Philip Rasmussen, led a trio of B-17Es on a long-range sector patrol. At 0300 they departed Koumac in northern New Caledonia and headed towards Guadalcanal. Lieutenant Robert Loder flew 41-9224 *Kai-O-Keleiwa* but after about an hour of flying on instruments, turbulence made it difficult to maintain control and formation in rain and poor visibility. Rasmussen abandoned the mission and turned back for Koumac. The airfield was socked in however, so Rasmussen headed for Plaine des Gaiacs, about 50 miles to the south-east. A radio operator there reported a weak signal on Loder's assigned frequency around the time he was due to run out fuel. The wreckage of *Kai-O-Keleiwa* has not been located on New Caledonia to date and it

Aircrew of an 11ᵗʰ BG Fortress clean their weapons on Efate in the New Hebrides. The weapon on the right is a 0.30-inch calibre M2 machine gun, and the larger one on the left is the barrel of a 0.50-inch calibre gun.

appears the bomber went into the sea. Thus, in just a week of operations, the 11ᵗʰ BG had lost four Fortresses and three crews.

Also active over the Solomons on 8 August was LB-30 AL573 which flew a fourteen-hour reconnaissance mission over the Solomons, one of eight such missions flown during the month. However, the small fleet of LB-30s was increasingly being used for transport missions, including to Milne Bay as will be seen in Chapter 10. By the start of September, the surviving LB-30s had been transferred to the 21ˢᵗ Troop Carrier Squadron.

On 6 August four *Kiyokawa Maru* Jakes had temporarily deployed to Tulagi. On this date they dropped 60-kilogram bombs on the Fo'ondo mission on the island of Malaita where six natives were killed. This was in relation to a small nearby Japanese communications station manned by two radio operators and 21 men which was wary of being observed by nearby native constabulary loyal to Allied coastwatchers still in the area. This task followed similar missions by *Kiyokawa Maru* Jakes on 8 and 31 July.

As mentioned in Chapter 7, in mid-July four No. 14 *Ku* Emily flying boats had arrived in Rabaul. These possessed a much longer endurance than the existing Mavises and soon were used for the raids in late July against northern Queensland targets including Townsville. From Tulagi the Yokohama *Ku* Mavises had also ranged as far afield as Efate, some 800 miles to the south, where airfield construction was confirmed in June.

On the morning of 3 August two No. 14 *Ku* Emilys were detached to Tulagi from where they flew strategic reconnaissances of Fiji and New Caledonia. These landmark missions, largely ignored by history to date, were finalised by the crews in a temporary headquarters set up in Tulagi township that evening. The first Emily headed for New Caledonia at 2345 and did not return until 1620 the next afternoon of 4 August. The second Emily departed for Fiji at 0240 on 4 August and returned at 1715 that afternoon. These exhausting missions, one of which exceeded sixteen hours, turned in nil results due to poor visibility, however their ambitious objectives are hard to ignore.

Around this time the huge Allied invasion convoy, comprising 75 vessels and three aircraft carriers, was in the area. It had departed Fiji on 1 August after amphibious landing exercises were carried out. The convoy was south of the New Hebrides and only 40 miles from Efate on 3 August, before turning north for the Solomons. During the next three days the convoy was within range of the regular Mavis searches being flown from Tulagi and both Emily flights of 4 August as well. In normal conditions the massive collection of ships could hardly have avoided detection. However, luck was with the Allies, with bad weather and thick cloud shielding the convoy.

Subsequently the landings proceeded as planned at dawn on 7 August, with the surprise factor being a major reason for the success. The First Marine Division faced some tough opposition against their 3rd Kure SNLF counterparts on Tulagi, but little opposition was faced against the largely unarmed and civilian Formosan construction personnel on Guadalcanal where the airfield had just been declared ready for operations.

It is often remarked that the Guadalcanal airfield was due to be declared operational on the day of the invasion. Indeed, staff officers from Rabaul were due to fly to the site in a Mavis flying boat on 6 August to inspect the area. This visit was unexpectedly delayed for a day, a delay which no doubt saved the lives of all of those scheduled to make the trip. Nevertheless, this close-run aspect of the airfield becoming operational remains theoretical. Among the many urgent messages scurrying between Rabaul and Tokyo in this busy week was a critical one which informed a begrudging Rabaul of a limiting factor: the designated fighter unit intended to base itself at the new airfield, No. 6 *Ku*, could not deploy to the South Seas theatre before 16 August.

Meanwhile on 7 August the most senior Japanese officer at Tulagi was the Yokohama *Ku hikocho* Captain Miyazaki Shigetoshi, who personally raised the alarm with Rabaul and was killed the following day defending the base. Miyazaki declared on the radio that his garrison would resist to the end. Although armed with only light weapons, it nonetheless took two days to finally rout all of the Japanese from Tulagi and surrounds.

Among the first to be notified of Miyazaki's appalling predicament at Rabaul was Rear-Admiral Yamada Sadayoshi, the CO of the 25th Air Flotilla, whose headquarters were housed in a stately wooden building stemming from the Australian colonial era. Yamada was much invested in the unfolding proceedings for it had been largely his decision to establish the airfield at Lunga Point on Guadalcanal. Despite the best efforts of the 11th BG, construction had so far proceeded almost unhindered and the appearance of American ground troops, up until this juncture, had been unthinkable.

During dawn strikes by US carrier aircraft that morning the Yokohama *Ku* detachment at Tulagi was wiped out, with the entire Rufe contingent of about fourteen Rufes destroyed on the ground at Halavo. Post-war wreck surveys confirm that seven Mavis flying boats were also burned to destruction at their moorings just to the north-east of Tanambogo. Two Mavises in particular were unlucky, with their crews already aboard preparing for a 0700 departure to conduct regular morning searches. An unknown number of crew tried to swim for safety, but those who survived would have soon been killed by US Marines.

Despite being taken by surprise, the Japanese response to the landings was swift. Miyazaki's bewildering radio call from Tulagi interrupted Rabaul's staff officers preoccupied with the three immediate tasks: undertaking raids against the newly-discovered Milne Bay airfield, returning the Tainan *Ku*'s Lae detachment back to Rabaul, and organising the arrival of the composite No. 2 *Ku* at Rabaul. Thus, while aircraft engines were being warmed for a major Milne Bay bombing attack scheduled for that morning, Yamada was spurred to quickly change the target to Tulagi and "Runga" on Guadalcanal. With 27 No. 4 *Ku* Bettys already loaded at Vunakanau with a mixture of 60- and 250-kilogram bombs, there was insufficient time to change the payload to torpedoes, an exponentially more effective anti-shipping weapon.

While the Betty crews were briefed on their revised target, at Lakunai airfield Tainan *Ku hikotaicho* Nakajima Tadashi also convened his pilots. At the base of the verandah of the wooden operations shack they were told their new target was "Runga", a reference to Lunga Point on Guadalcanal. This was an unexpected and dramatic change which would encompass the longest combat range they had ever flown in their Zeros, it being approximately 600 miles from Rabaul to Guadalcanal. Among the many surreal scenes was one of the Tainan *Ku hikocho* Captain Saito Masahisa arguing with his superior, Admiral Yamada. This flag officer urged the launch of all two dozen Tainan *Ku* Zeros already readied for the Milne Bay mission. Saito, who appreciated the practical dangers and limitations associated with such a long-range mission, persuaded his superior to reduce the number to eighteen, a decision immediately relayed to *hikotaicho* Nakajima.

While Saito and Yamada robustly conferred, Nakajima had concocted a leadership structure for the exceptional mission which defied normal IJN adherence to seniority, and instead emphasised reliance on experience. In an exemplar example of IJN flexibility under duress, Nakajima would lead the mission himself and placed one of the unit's higher-scoring pilots, FPO1c Nishizawa Hiroyoshi, as his left-land wingman. Nakajima's Zero was a *hokoku* fighter; one donated by a patriotic group. Only a few were spread throughout the Tainan *Ku* inventory, and it seemed fitting that the *hikotaicho* would choose such an aircraft to showcase Japanese patriotism. The fighter was tail-code V-138 and had a double blue fuselage sash alongside *kanji* characters on the rear fuselage proclaiming it had been donated by the Haysahi Forging Company headquartered in Hebei, Japan.

Nakajima also restructured his strike force so that his most senior and experienced *buntaicho*, Lieutenant Kawai Shiro, would fly ahead of the others for the purpose of "air control", meaning Kawai was placed to find trouble first. Nakajima's other two *chutai*, led by himself and Lieutenant Sasai Jun'ichi, would fly close protection for the 27 Bettys led by Captain Egawa Renpei. Nakajima emphasised at the briefing that the role of bomber protection was paramount, and was to be carried out at all costs.

Seated in the observer's position of the lead bomber, Captain Egawa Renpei had only recently been promoted to Captain. When the US declared war on Japan he had been the Air Officer aboard the cruiser *Isuzu*, then in May 1942 he was promoted and transferred to the light cruiser *Katori* as Division Officer. He had only arrived in Rabaul some two months ago and was still familiarising himself with land-based operations. Egawa's 27 Bettys lifted from Vunakanau at 1006 and formed up with Nakajima's Zeros over the ocean south-east of Simpson Harbour. The forthcoming battle harshly dealt with the formation: six Bettys were lost, fifteen more were hit by gunfire and 29 crew were killed. In addition, two Zeros failed to return, while the famed ace FPO1c Sakai Saburo was wounded in the head and survived an epic flight back to Rabaul. He was later medically evacuated by Mavis back to Japan but would never again fly in the South Pacific.

Following this was a desperate attack by a *chutai* of nine Vals from the newly arrived No. 2 *Ku*. These did not have the range to reach the target and return to airfields, instead the forlorn plan was for them to ditch at the seaplane base in the Shortlands on their return leg. All of the Vals were lost, and just a handful of airmen were rescued by a destroyer after ditching. The results of the day's Japanese aerial counterblow were close to negligible: a USN transport was hit that later sunk and a destroyer was damaged.

The naval action taken by the Eighth Fleet commander, Vice Admiral Mikawa Gun'ichi, is worthy of mention. Mikawa assembled a force of five heavy cruisers, two light cruisers and a destroyer to make a night attack on the Allied shipping off Guadalcanal. Mikawa's force was spotted by RAAF Hudsons from Milne Bay when east of Bougainville on 8 August, but for various reasons these reports never properly alerted the Allied naval forces and an associated long-running controversy is discussed further in Chapter 10. That night Mikawa's force scored a decisive victory and sunk four Allied cruisers at the Battle of Savo Island. Among other things, this underscored Japanese naval strength in the region. With Allied warships concentrated in the Solomons, subsequent Allied operations at Milne Bay were undertaken with virtually no significant naval support with which to counter Japanese moves.

For months the brunt of Japanese air operations in the South Pacific had been borne by two hard worked units: the No. 4 *Ku* and the Tainan *Ku*. However, neither was strong enough to wage both offensive and defensive operations over several parts of New Guinea and the Solomons. Fortunately for the Japanese, the 26th Air Flotilla which had been earmarked to occupy the newly completed Guadalcanal airfield was immediately available. As has already been mentioned, the No. 2 *Ku* had already arrived at Rabaul with its mix of fighters and dive-bombers. Joining that was the Misawa *Ku* whose 27 Bettys arrived from the Central Pacific on 8-9 August. Two more reinforcement units would arrive later in August. The first was the Zero-equipped No. 6 *Ku*, originally slated to be based at Guadalcanal, and intended to arrive at Rabaul by 16 August. At this juncture the date was set back even further, however, and the units' training program in Japan was accelerated so that it would arrive in Rabaul on 21 August. The second of these units was the Kisarazu *Ku*, which operated Bettys with the tail-code prefix R. It had been reorganised as a combat unit in Japan on 1 April 1942 and the first *chutai* flew into Vunakanau from the Marianas on 23 August.

The heavy attrition suffered by the Japanese attackers during the 7 August 1942 missions set the new standard for Solomons operations in the weeks and months to come. It is at this juncture that the Guadalcanal fighting becomes effectively a new and distinct theatre, separate to New Guinea. Thus, this chronicle now ends its detailed coverage of Solomons operations by both sides. However regular references will continue to the theatre, especially pertaining to losses sustained by the air forces at Rabaul. The Japanese there now faced combat on two additional fronts after two major shocks in the first week of August: the discovery of the Allied base at Milne Bay quickly followed by the American invasion of Tulagi and Guadalcanal. From now on "Rabi" and "Runga" would feature regularly as destinations in Rabaul's operations records.

A Betty bomber sinks in the ocean off Tulagi on 7 August, one of six lost during a costly series of missions in response to the American landings in the Solomons.

The USN heavy cruiser USS Astoria seen off Guadalcanal on 8 August. That night she was one of four Allied cruisers sunk during the Battle of Savo Island.

No. 76 Squadron's first aerial victory occurred on 4 August when pilot FPO2c Hanahiro Keiryu and observer WO Hasegawa Kameichi were shot down in their C5M Babs while conducting a reconnaissance of "Rabi", the Japanese name for Milne Bay.

CHAPTER 9

PRELUDE
MILNE BAY 4–22 AUGUST

As noted in Chapter 3, the first convoy arrived at Milne Bay on 25 June and construction activities were soon underway for a major new base there. Milne Bay itself was a significant landmark, being seven miles wide at its seaward end and extending westwards for twenty miles. It was at the head of the bay where a coconut plantation on a coastal plain was a suitable site for an airfield, although nearby ridges rose to 5,000-feet just a few miles inland. With the short voyage from Port Moresby taking less than two days sailing, soon a regular shuttle of ships was arriving and unloading supplies at the head of the bay near a village called Gili Gili. Just a few miles away was the village of Rabi, from where the Japanese drew their name for the location. In Allied communications the site was often referred to by its codename of Fall River.

The construction activities at Milne Bay were among the more successful Allied achievements in the South Pacific in 1942. Within weeks a wharf had been constructed as well as two runways and a garrison for thousands of troops. All this was done under strict secrecy and radio silence such that the Japanese did not discover the site until construction was well advanced. This was on 3 August when a Betty on a reconnaissance mission overflew the area, as noted in Chapter 8.

Unfortunately, Milne Bay suffered from an extraordinary amount of rain (over 100 inches annually) such that much of the area was a muddy and boggy morass. Construction of the first 5,000-foot runway, known as the No. 1 Strip, was quickly completed but the surface was often covered with water and operations were only possible due to panels of interlocking steel matting laid down to a width of 80-feet. Referred to as Pierced Steel Planking (PSP), it was also known to the Americans as Marston matting from the town in South Carolina where it was made (frequently misspelled as Marsden matting). Even so, landing on such a surface was hazardous and aircraft were prone to skidding off the runway, often as a result of aqua-planing. At first there was limited steel matting available for use on taxiways and dispersal areas, so bogging in these locations was a constant risk. Nonetheless, after four weeks No. 1 Strip was largely completed, and the first RAAF Kittyhawk landed there on 19 July. Such was the intended scope of operations from Milne Bay that a site for No. 2 Strip was already being cleared at Waigani, a few miles inland from No. 1 Strip, that would soon be available for emergency use (not to be confused with similarly-named Waigani near Port Moresby). Another coastal location to the east for a third strip was already being surveyed.

As outlined in Chapter 2, the two Kittyhawk-equipped RAAF fighter squadrons, Nos. 75 and 76, had been training and re-equipping in Queensland and would be moved to Milne Bay to provide for the air defence of the new base. The ground contingent of No. 76 Squadron left Townsville by ship on 14 July and arrived at Milne Bay four days later. On 19 July the unit's

Kittyhawks departed for the ferry flight to Port Moresby, where the fighters were held pending completion of dispersal pens at Milne Bay. While at this location seven No. 76 Squadron Kittyhawks were ordered to bomb enemy positions at Gona on 22 July, but instead skirmished with Tainan *Ku* Zeros as detailed in Chapter 6. Three days later the squadron's Kittyhawks flew into Milne Bay.

Meanwhile an advance party from No. 75 Squadron arrived at Milne Bay via an Empire flying boat on 22 July, with the remainder of the squadron following by ship on 30 July. By the start of August both RAAF fighter squadrons were operational at the new base, as evidenced by the Betty reconnaissance of 3 August which counted 39 P-40s on the runway, dispersal areas and in the air. However high rainfall was only one of several factors forbidding to the conduct of safe visual flying operations. The persistent low cloud base, compounded by mountainous terrain a few miles inland from the airfields, made any aerial activity in marginal conditions dangerous indeed. Also, just because it was deemed safe to depart the area in low cloud did not mean it was safe to return. Meteorological conditions could quickly and dramatically worsen. This was particularly dangerous for fighters returning with marginal fuel reserves, as we shall see. All of this meant that aerial operations were more often curtailed by the weather, rather than the enemy.

The overall conditions were summed up by a meteorological officer who was stationed at Milne Bay in 1944-45 who stated that it was:

> ... possibly one of the most forbidding spots on earth ... characterised by the worst weather conditions in my experience.

Following the Betty mission on 3 August, a Tainan *Ku* reconnaissance of Milne Bay was urgently ordered for the following day. This was by a C5M Babs, piloted by FPO2c Hanahiro Keiryu with observer Warrant Officer Hasegawa Kameichi, with an escort provided by four Zeros. The aircraft departed Rabaul at 0930 and arrived over the target area at 1305 where they ran into a patrol of eight Kittyhawks. As fighter to fighter combat broke out in cloudy conditions, some minutes later two Kittyhawks piloted by Flight Lieutenant PH Ash (A29-90) and Sergeant Gray (A29-75) came upon Hasegawa's Babs a good thousand feet distant from the escorting Zeros.

Ash was credited with shooting down a "dive-bomber" after he reported that his gunfire produced heavy smoke from his victim and that it had a fixed undercarriage. Ash's aircraft was slightly damaged in the engagement, but no other aircraft were lost despite additional kill claims by both sides. The wreckage of the Babs was found by Australian soldiers three days later and a more detailed examination subsequently confirmed it was a land-based reconnaissance plane. The incident marked No. 76 Squadron's very first confirmed kill.

Rabaul's IJN high command had placed much weight on obtaining photos of the new Allied airfields under construction. When the Babs failed to return, along with the invaluable photography it was tasked to produce, Rabaul assumed the worst: that the infrastructure there was truly menacing to Japanese objectives. Accordingly, plans were put in place to commence an aerial campaign against the remote location.

Being 300 miles east of Port Moresby, the great advantage of Milne Bay to the Allies lay in its

An RAAF Kittyhawk on a narrow Marston matting taxiway at Milne Bay. The airstrips and base area were cleared from a coconut plantation which can be seen in the background.

ability to project sustained reconnaissance cover over the Solomon Sea towards New Britain, Bougainville and beyond. To this end on 5 August five No. 32 Squadron Hudsons arrived at Milne Bay from Horn Island. These were under the command of Flight Lieutenant Lloyd Milne, a veteran of No. 24 Squadron's Rabaul detachment at the start of the war. Soon the aircraft were flying regular six-hour reconnaissance patrols over enemy territory, with increased urgency and priority due to the upcoming Solomons operations.

On the morning of 8 August, two Hudsons sighted Mikawa's force of warships east of Bougainville which that night would sink four Allied cruisers during the Battle of Savo Island. One of the Hudsons, A16-218 flown by Sergeant WJ Stutt, reported the presence of a "Zero float plane" and escaped into clouds. The floatplane was likely a Jake observation aircraft from one of the cruisers. Stutt's radio operator sent an accurate description of the naval force, reporting eight warships present, but it was never acknowledged by Milne Bay due to an air raid alert at the base. Another Hudson, A16-185 flown by Flying Officer M Willman, also spotted the force but the radio operator made no transmission, wrongly believing he was under orders to maintain radio silence.

Nevertheless, by early afternoon both Hudsons had landed at Milne Bay and after verbal debriefings of both crews, both sighting reports were transmitted. However, a misinterpretation arose when the inclusion of two seaplane tenders somehow led intelligence analysts to assess that the ships were primarily involved with setting up a seaplane base. For this reason, further reconnaissance flights of the area were not launched which would likely have confirmed the warships continuing their southerly course towards the lower Solomons. For obscure reasons, further delays in processing these reports meant that they were not received in time to alert the naval forces off Guadalcanal, and within hours the disastrous Battle of Savo Island commenced.

Years later, the American naval historian Samuel E Morison wrote his prize-winning fifteen volume *History of United States Naval Operations in World war II*. Morison blamed the failure of the timely transmission of the reports on a Hudson pilot who:

> ... spent most of the afternoon completing his search mission, came down at Milne Bay, had his tea, and then reported the contact.

This assertion has been repeated by many authors since, however it is unwarranted to pass blame on any single crew for the alleged failure; for a start USN sources do not agree on exactly when the report was eventually received. As with so many outcomes during the Pacific War, a myriad of factors was at play, including communications complications stemming from the Army / Navy command boundary between MacArthur's SWPA and the USN's SoPA command.

Meanwhile, after the initial Japanese appearances over Milne Bay on 3 and 4 August a USN radar station became operational which gave some warning of incoming raids, although successful interceptions were always problematic in the face of the unpredictable weather. On 7 August No. 76 Squadron was ordered to scramble on account of a warning, but no contact resulted. However, five inexperienced pilots remained airborne too long searching for the enemy such that they were forced to land at an emergency strip on Goodenough Island, some 60 miles from Milne Bay. One of the Kittyhawks, A29-81, was written off while landing. The stranded pilots were supplied via air-dropped rations until arrangements were made to fly out the surviving four machines.

As already mentioned, at Rabaul on the morning of 7 August the Japanese had readied a strike force of Bettys and Zeros to attack Milne Bay. However, following news of the Guadalcanal landings, this force was directed to attack shipping there instead. Despite grievous losses of 17 Bettys over Guadalcanal on 7-8 August, an attack on Milne Bay was ordered for 11 August.

This was carefully planned to be a multi-wave affair. The main strike would be undertaken by a dozen No. 4 *Ku* Bettys led by *hikotaicho* Lieutenant-Commander Mitsui Koji. The bombers would be escorted by fifteen Tainan *Ku* Zeros led by *hikotaicho* Lieutenant-Commander Nakajima Tadashi, which had the secondary task of strafing the airfield. This combined formation would be preceded by seven No. 2 *Ku* Model 32 Zeros led by Lieutenant Kurakane Yoshio which would sweep over Milne Bay half an hour earlier.

The presence of two such senior *hikotaicho* underlines the importance which Rabaul attached to this inaugural mission, but things went awry from the outset. The same weather which so often dogged Allied aerial operations now frustrated the best Japanese efforts. After launching at 0900, Kurakane's Zeros battled persistent low cloud and squalls which hid most of New Guinea's eastern tip. After stooping around to find a gap in the weather they returned five hours later to Lakunai.

Kurakane's Zeros were followed at 0930 by the main combined formation, but Nakajima quickly aborted due to a faulty oxygen regulator. Command of the Zeros was handed over to *buntaicho* Lieutenant Sasai Junichi, marking the only occasion Nakajima ever aborted a combat mission in New Guinea. However, the combined formation soon entered the same capricious weather which Kurakane was negotiating and at 1040 Mitsui ordered his bombers to turn back, all of which landed safely at Vunakanau early that afternoon. Using hand signals, Sasai ordered seven Zeros to escort the bombers home, leaving his own *chutai* of six Zeros to persist ahead.

Sasai guided his six Model 21s through thick cloudbanks due east of Milne Bay where they found Samarai at 1150. Throttling back to save fuel, Sasai scoured the marginal visibility for

P-40E-1 Kittyhawk A29-125, Y-WURRY, No. 75 Squadron, RAAF, Milne Bay, August 1942.

the target. Then, almost an hour later at 1245, they suddenly ran head-on into a combined formation of sixteen RAAF Nos. 75 and 76 Squadron Kittyhawks. In fact, Sasai's formation had been detected by radar and as they searched for a way through the cloud this gave ample time for the Kittyhawks to scramble. Nonetheless, the precipitous encounter surprised both parties.

Widespread and confused combat now unfolded between 1,000 to 3,000 feet, but Sasai's experienced team was able to pounce on mistakes by several inexperienced Australians, and three No. 75 Squadron pilots were downed and killed. Flying Officer Mark Sheldon in A29-123 was shot down into the side of a mountain on the northern side of the bay, approximately 25 miles east of Gilli Gilli.   A search party identified his wreck the following day. Warrant Officer Francis Shelley (A29-100) and Flying Officer Albert McLeod (A29-93) disappeared somewhere into Milne Bay's cavernous geography (McLeod's fighter was found in 1967).

No. 76 Squadron also had its share of disaster. At the end of the battle, Flight Sergeant George Inkster flew his damaged Kittyhawk low over No. 1 Strip accompanied by excited radio chatter which preceded the engine bursting into flames. Inkster bailed out, but with a flailing unopened parachute he fell to his death, perversely just ahead of the squadron ambulance. Overall, the two RAAF squadrons had lost four Kittyhawks and their pilots. To complete the fragmented ledger, there were still five machines marooned on Goodenough Island.

On the Japanese side, no Zeros were lost in combat although FPO3c Endo Masuaki was seriously wounded by Kittyhawk gunfire. He ditched near Buna among friendly ships at 1240 rather than facing an extended and possibly fatal attempted return to Lakunai. Endo was returned to Rabaul by ship and admitted to the well-equipped IJN hospital there where he convalesced for several weeks before being returned to duty.

There was now a lull in enemy operations over Milne Bay for almost two weeks, mainly as Rabaul was focused on the energised Guadalcanal campaign. However, during this period Allied Intelligence received forewarning of a new Japanese land offensive that would take place between 22-27 August. At the time the precise target was unknown, but Milne Bay was an obvious possibility and the entire garrison was put on high alert. Air patrols by the Kittyhawks were increased, and logistics were set in place for them to be used as fighter-bombers when necessary.

High rainfall and poor drainage at Milne Bay led to extremely challenging conditions for men and machines. This Australian army truck is hopelessly bogged near Gili Gili in October 1942.

Supply ships unloading at Milne Bay in September 1942. The construction of a major new base over a period of several weeks in June and July was a major Allied accomplishment.

Since the development of the Milne Bay base the site had been well served by regular transport flights, including those by No. 33 Squadron Empire flying boats for which the sheltered waters at the head of the bay formed a natural alighting area. On 18 August LB-30 AL515 *Yard Bird* was used to deliver a Bofors gun and crew to Milne Bay from Townsville. This mission was repeated two days later, but a hydraulic breakdown meant the landing gear would not lower and lock. After struggling for some time with the manual release procedure, the starboard gear would not let down and the crew crash-landed under a very low cloud ceiling and failing light at 1745:

> We made two low passes over the landing strip – every Army and Air Force bod from the area had lined the strip. We made a long low approach and touched down on one wheel on the muddy strip beside the steel runway. As we lost speed we went down on our nose and the starboard wingtip and two props touched. At 90mph we made a belly skid to the right, finishing up the way we had approached.

Crews from the RAAF fighter squadrons stripped the Liberator of its guns and ammunition, but the derelict airframe was too bulky to move and it remained a prominent sight, tipped to one side on the No. 1 Strip. The LB-30 crew was flown back to Townsville by an Empire Flying Boat on 25 August.

A significant Allied strategic decision was made on 19 August, the day after the LB-30 accident. As will be noted in Chapter 11, on 9 August B-26 pilot Captain Winifred Craft had landed his bomber at Milne Bay airfield and assessed it as suitable for B-26 operations. Since then, the weather had become an increasingly detrimental factor at Milne Bay and the 22nd BG commander, Lieutenant–Colonel Dwight Divine had become aware of RAAF difficulties in operating their Kittyhawks on the wet runway. Nonetheless, B-26 operations would potentially become much more effective and wide-ranging were they be able to use Milne Bay as a forward base.

With these considerations in mind, Divine set off from Port Moresby on the morning of 19 August in a B-26 with 22nd BG operations officer Major Hugh Manson to personally inspect the facilities. Their first attempt to get through was thwarted by a low ceiling frontal weather system, and Divine was soon back at Seven-Mile. He landed his B-26 on Milne Bay's wet PSP on his second attempt later that afternoon whereupon he both inspected the area and held discussions with RAAF operations officers. The ongoing weather was atrocious, there were inadequate dispersal areas, and the PSP runway was particularly slippery for heavier aircraft. There and then Divine decided Milne Bay was unsuitable. Upon return to Seven-Mile he recommended to the US bomber commander, Brigadier-General Kenneth Walker, that the field was unsuitable for USAAF medium bombers. As a result, no B-25s or B-26s would ever use Milne Bay as a base, and it became a centre almost exclusively for RAAF operations.

At this time the relief of the hard-worked Hudsons of No. 32 Squadron began, as the squadron had been operating from Port Moresby and Horn Island continuously since February. Accordingly, fifteen Hudsons from No. 6 Squadron started the flight from Richmond, New South Wales, to Horn Island on 21 August. However operational needs would mean the relief was not clear-cut, and various elements of both squadrons would operate concurrently for several days until No. 32 Squadron was fully withdrawn to Richmond by early September.

During this period of flux an accident on 16 August underlined the operational risks taken on a daily basis at Milne Bay even in the absence of the enemy. Flying No. 76 Squadron Kittyhawk A29-78 (coded "IR" and named *Bloody Mary*), Flying Officer Colin Lindeman experienced a tyre burst on take-off. The fighter veered off the Marston matting and collided with No. 32 Squadron Hudson A16-218, writing off both aircraft and killing Sergeant Patrick Ellis.

A No. 75 Squadron Kittyhawk and Tainan Ku Zero engage over Milne Bay in August 1942.

CHAPTER 10

BUNA RESUPPLIED
NEW GUINEA 9–22 AUGUST

In the early hours of 9 August a pair of Catalinas, A24-18 and A24-27, continued the series of night nuisance raids over Lakunai and Vunakanau airfields, dropping bombs on the runway areas from 8,500 – 10,000 feet. The effort was repeated the following morning by A24-26, and amid clear conditions the crew claimed a bomber destroyed and large fires started. Two other Catalinas were not so lucky. A24-25 experienced bad weather over Lae and had to jettison its bomb load, while A24-1 experienced engine trouble on the way to Rabaul and turned back to base after dropping a bomb over the Gona-Kokoda area. In the early hours of 11 August A24-12 made another raid on Lakunai and noted some 40 ships in Rabaul harbour.

On 9 August the 19th BG continued its effort to subdue Rabaul, although at a lesser extent than the sixteen B-17s put over the town two days earlier. This time five Fortresses departed from Seven-Mile, although one suffering from an ailing engine split off to bomb the secondary target of Gasmata instead. This was an afternoon mission, timed to arrive over the target just before dusk. It was led by Lieutenant Clyde Kelsay in *Spawn of Hell*, and comprised Fortresses from the 28th, 30th and 93rd Bombardment Squadrons.

The quartet of B-17s dropped their bombs on Lakunai but were then intercepted by fourteen Tainan *Ku* Zeros led by the experienced *buntaicho* Lieutenant Kawai Shiro. The Zeros had been forewarned of the approaching bombers which enabled them to climb their altitude and make the interception. In the last rays of the sun at 1750 the Zeros made concerted attacks on the formation and fierce combat took place. American gunners claimed seven Zeros shot down, but none received any damage. The Zeros, airborne for exactly one hour, claimed a Fortress.

One of the two 93rd BS B-17Es in the formation was unnamed 41-2643 flown by Lieutenant Hugh Grundmann. It approached the target positioned on the far right of the formation but lagged behind the other three due to rough engines unrelated to combat damage. Quickly the Zeros turned their exclusive attention to the straggler, whose demise was not seen by the other Fortresses. The crew was posted as missing when the aircraft failed to return, however post-war examination of the wreckage in 1946 indicates it dived into the jungle near Keravat, killing all aboard instantly. The scattered wreckage lay in thick jungle near Rondahl's Plantation where a local missionary had buried two of the bodies. The remaining crew were recovered post-war from nearby burial sites, likely buried by the Japanese.

As the surviving three Fortresses made the return flight to Seven-Mile they separated in the dark. Captain Harry Hawthorne flying the unnamed 93rd BS Fortress 41-2452 was soon in difficulty as two engines had been damaged along with his radio compass. The impaired engines had slowed the aircraft under the economical cruise speed and increased fuel consumption.

Tainan Ku Buntaicho Lieutenant Kawai Shiro who adopted innovative new interception tactics to combat B-17 raids over Rabaul. Kawai aggressively led from the front, and it was a rare day when he did not fly.

Lost and unable to get radio bearings from Port Moresby, after three hours of searching unsuccessfully Hawthorne decided he would have to ditch. He put the Fortress down just offshore Malapla Island, not far from Milne Bay. The crew was soon rescued by an Australian naval vessel and returned to Port Moresby.

Also on 9 August nine B-26s, mostly from the 33rd BS, made a low-level raid on Salamaua in the mid-afternoon during which gunners also strafed the town. This was led by Captain Winifred Craft. The following morning all the Marauders returned to Townsville, except Craft. Flying a bomber named *Martin's Miscarriage*, Craft took the opportunity to make the first B-26 landing at Milne Bay and reported the runway as fine for operations, a decision that was soon reversed as described in Chapter 10.

After his success at the Battle of Savo Island, Vice-Admiral Mikawa returned to Rabaul where he returned the most powerful part of his force, the four heavy cruisers of the Sixth Cruiser Division, to the anchorage at Kavieng. After zigzagging as an anti-submarine manoeuvre while in the natural choke point of St George's Channel between New Britain and New Ireland, the ships emerged in open water on a bright sunny morning of 10 August.

With the ships steaming in line astern at 16 knots, and attack believed to be unlikely, the skippers ordered the portholes opened to give some relief to the stifling heat below decks. This was welcomed by the crewmen who had been locked down for combat for the best part of the last 48 hours. However, while less than 100 miles from Kavieng, at 0808, the last cruiser in line astern, *Kako*, was struck by three huge explosions. The bow of the heavy cruiser dipped and twisted below the waterline letting water flow in through the open portholes. Within minutes *Kako* had sunk, taking 68 crewmen to their death. Another 649 were rescued by the other three cruisers. The culprit was USN submarine *S-44*, which in one stroke had partly avenged the Allied losses at Savo Island and following which it returned safely to Brisbane on 23 August.

Bad weather descended on the Port Moresby area during 10 and 11 August, which limited air operations, with the critically needed transport flights to Wau and Myola resuming on 12 August. On the same day the 19th BG resumed its Rabaul campaign, with a raid by seven Fortresses from the 28th, 30th and 93rd Bombardment Squadrons. The raid targeted the harbour, which was crowded with shipping of all kinds at this time, as reported by a Catalina on a night mission noted above which reported 40 vessels there early the previous morning.

The raid was led by Captain Hillhouse of the 30th BS, with 48 x 500-pound bombs dropped from 26,000 feet. The Americans claimed to have hit three transports, but only the *Matsumoto*

Maru received minor damage. The Fortresses were intercepted by fifteen Tainan *Ku* Zeros, again led by stalwart Lieutenant Kawai Shiro.

The Fortress raids were increasingly concerning Rabaul high command and different options had been considered in how best to deal with them. Kawai was by far the most experienced Zero combat leader at Rabaul at this juncture of the war, and he led most Tainan *Ku* defensive missions over Rabaul. For defensive fighter strategy, Kawai had replaced the Tainan *Ku*'s traditional *chutai* combat structure with smaller formations of individual *shotai*. For this particular mission he had structured the Zeros into five *shotai*, each with three aircraft. Kawai had also experimented with two-fighter and four-fighter *shotai* depending on the experience level of the relevant *shotaicho*. Another factor was that an increasing number of officer pilots had arrived in the unit in July / August, so the new structure also offered ample opportunity for leadership experience. Several bombers received slight damage after Hillhouse maintained a tight defensive formation. Despite the Fortress' defensive fire, not one Zero sustained a single bullet hit. Kawai's Zeros, wary of the possibility of more Fortresses appearing for a second raid, stayed aloft over Rabaul for another three hours.

The shipping concentration at Rabaul was partly connected with renewed attempts to send a new convoy to the Buna beachhead area. This was the fifth Buna convoy, as detailed below:

Fifth Buna Convoy

- *Tatsuta* (4,350-ton light cruiser; built 1919; 4 x 5.5-inch guns; 6 x torpedoes)
- *Yuzuki* (1,772-ton destroyer; built 1927; 4 x 4.7-inch guns; 6 torpedoes)
- *Uzuki* (1,772-ton destroyer; built 1926; 4 x 4.7-inch guns; 6 torpedoes)
- *CH-23* (438-ton submarine chaser; built 1941; 1 x 3-inch gun)
- *CH-30* (420-ton submarine chaser; built 1942; 1 x 3-inch gun)
- *Kinai Maru* (5,040-ton IJN auxiliary transport; built 1930)
- *Kenyo Maru* (6,486-ton IJN auxiliary transport; built 1938)
- *Nankai Maru* (5,114-ton IJN auxiliary transport; built 1933)

The three transports in this convoy were loaded with 3,000 men of the 14th and 15th Establishment Units, together with construction equipment, vehicles and 70 tons of IJA supplies. The destination was Buna so that the airfield could be readied to receive aircraft. The convoy had originally departed Rabaul on 6 August but following the Guadalcanal landings it was recalled by a cautious Mikawa and returned to Rabaul on 9 August.

The convoy departed once again on 12 August, not spotted by Allied reconnaissance until the morning of 13 August as it approached the Papuan coast. On this same day the *Hikotaicho* of the Tainan *Ku*, Lieutenant-Commander Nakajima Tadashi, decided to forward-deploy sixteen Zeros to Lae to cover the convoy and the Buna area. This was a difficult decision to make, as Nakajima's fighters now faced threats on multiple fronts – Rabaul, Guadalcanal, Milne Bay and the Buna-Lae beachhead area.

Accordingly, on 13 August a series of Tainan *Ku* patrols from both Rabaul and Lae provided the ships air cover as they approached Buna. During one of these missions Lieutenant Murata Isao was lost with his fighter to unspecified operational causes at Lae, possibly a take-off or landing accident on Lae's badly damaged runway. A force of ten Fortresses was launched from Port Moresby at 0840, but only seven made it to the beachhead area where they ran into heavy cloud and rain. From 16,000-feet 24 x 500-pound bombs were dropped but results were not observed due to the conditions. At 1007 one of these Fortresses was observed through cloud breaks by the first patrol of three Zeros from Rabaul, however they were unable to engage the bomber due to the weather.

A strike by five B-26s was launched that afternoon, with escort provided by 80th FS P-400s. Over the target area these ran into a protective umbrella of eight Zeros led by Lieutenant Yamashita Joji. This Zero formation was one of the Lae-based patrols which attacked the B-26s at 1520. Two of the bombers were damaged, with Lieutenant Harry Patteson being forced to ditch *Sally Rand* after the starboard engine was shot away. The bombardier, Lieutenant Duncan Hughes, died of wounds sustained on impact. The remainder of the crew were rescued from Porlock Harbour on the northern Papuan coast by an RAAF Catalina the following day.

Other B-17s attempted to find the ships shortly before dusk, with Major Rouse leading a trio that found a lone cruiser, the *Tatsuta*:

I got through OK with my flight and bombed a cruiser. He manoeuvred and we missed.

To Rouse's frustration he then flew over Buna and spotted the three transports which presented relatively easy targets, but all of his bombs had been expended.

With Rabaul's' Zero inventory much reduced with the temporary forward-deployment to Lae, Nakajima placed the Tainan *Ku*'s C5M Babs on increased aerial patrol duties. On 13 August a pair of Babs commanded by Lieutenant Kizuka Shigenaga ran a series of four overlapping patrols over Rabaul from 1040 onwards with no encounters.

At dawn on 14 August a lone 435th BS B-17E named *Chief of Seattle* departed Port Moresby for one of the regular daily reconnaissance flights covering Gasmata, Rabaul and Kavieng. The Fortress was so named after being paid for by the sale of war bonds by the citizens of Seattle and had only arrived in Australia on 2 August. With the crew under strict orders to maintain radio silence, it was never heard from again, with the loss of Lieutenant Wilson Cook and his crew.

After departing Port Moresby the Fortress crossed the Owen Stanleys and was due to over fly Buna on the way to Gasmata. It was here that it ran into a nine-strong Zero *chutai* led by Lieutenant Yamashita Joji which had just arrived in the area from Lae. The Zeros claimed the Fortress, which likely crashed into the ocean as no trace of it has ever been found. After not downing one of the big bombers for many months, the Tainan *Ku* had now honed their tactics and were downing B-17s with increasing regularity.

Return fire from the B-17 put a large gash in Lieutenant Ono Takeyoshi's starboard wing. Escorted by his wingman, Ono made a precautionary landing at Buna, and the pair became the very first Japanese aircraft to land there.

Lieutenant Wilson Cook and the crew of 435ᵗʰ BS B-17E Chief of Seattle at Seven-Mile in mid-1942. All were lost when shot down by Tainan Ku Zeros on 14 August 1942.

Meanwhile, seven B-17s had departed Port Moresby some 30 minutes after *Chief of Seattle*, led by Major Dean Hoevet the CO of the 30ᵗʰ BS. When these arrived over the beachhead area there was no sign of the ships, which had finished unloading during the night and were already on their return voyage to Rabaul. Hoevet began a search and the ships were eventually found amid poor weather some 100 miles north-east of Buna.

As Hoevet began his bomb run, the Fortresses were flying beneath low cloud at just 3,800-feet when they were attacked by Yamashita's remaining seven Zeros. On this occasion the Zeros succeeded in driving the B-17s away and protecting the convoy. However, FPO3c Arai Masami was shot down by the Fortresses' defensive fire, and running short of fuel most of Yamashita's pilots put down at Salamaua at 1050. Having lost contact with the ships, Hoevet's men had to settle for unloading their bombs over the beachhead area.

This incident was soon followed by a misfortune in Australia where on 16 August, Hoevet was lost in a tragic accident. After taking off from Mareeba in B-17E 41-2434 with eleven others aboard, Hoevet flew out to sea to conduct tests of a new flare-dropping mechanism. However, the drive mechanism faltered and one of the flares exploded inside the bomber, setting it on fire and causing it to crash into the ocean not far from Cairns. The airfield at Mareeba was subsequently named Hoevet Field.

A6M3 Model 32 Zero V-176, CN 3017, Tainan Ku, Chutaicho Lieutenant Ono Takeyoshi who made an emergency landing at Buna 14 August 1942.

On 17 August the Japanese resumed their attacks on Port Moresby, with the first daylight raid for over three weeks. The bombing force was significant, comprising nine No. 4 *Ku* Bettys and sixteen from the recently arrived Misawa *Ku*, led by the Misawa *Ku buntaicho* Lieutenant Nakamura Tomo'o. These were escorted by a mixed formation of fighters, a dozen Tainan *Ku* Zeros together with ten from the No. 2 *Ku*.

On this occasion the tactic used by Nakamura to approach the target area from seawards paid big dividends. Largely for this reason no advance warning at Port Moresby was received, and if the formation was detected on radar it was likely mistaken for Allied aircraft flying in from Queensland. Hence just four minutes warning was received, meaning that the defending Airacobras had no hope of making contact with the intruders. Port Moresby's AA batteries opened fire and five of the No. 4 *Ku* Bettys received shrapnel damage, but it was not enough to deter the bombing accuracy which was deadly accurate.

At 0915 dozens of 60- and 250-kilogram bombs tumbled earthwards and landed accurately on the Seven-Mile runway, where personnel were caught completely by surprise. Several aircraft were neatly lined up in the open including transports waiting to be loaded and a flight of B-26s warming their engines. From high above the Betty crews claimed "excellent results" and that six aircraft were destroyed.

The Japanese claim was reasonably accurate. One of the B-26s, *Shamrock*, was blown to pieces by a direct hit while another named *The Avenger* was seriously damaged while it was taxying and one crewman was killed. Of more significance was the damage done to aircraft of the hard-working and invaluable transport fleet, three of which were destroyed: DC-5 callsign VHCXA and two Lodestars (VHCAG and VHCAI). Four other transport aircraft suffered shrapnel damage and had to be flown to Australia for repairs: a DC-3 (VHCXG), two C-53s (VHCCB and VHCCC) and a DC-2 (VHCXG).

On the return flight to Rabaul the Japanese formation ran into a severe weather front, such that many aircraft diverted to Lae rather than risk flying through it. After months of experience with New Guinea weather just two Tainan *Ku* pilots chose to fly direct to Rabaul, and one of these, FPO2c Norio Tokushige disappeared without a trace.

The damage and destruction to transport aircraft had huge ramifications for the land campaigns which had become highly dependent on air transport of supplies into the mountains. Immediately after the raid just one Douglas transport aircraft was on hand, and appeals for further aircraft were met with the fact that only 30 transport aircraft were then available in the entire SWPA, with less than half available at any one time due to serviceability issues. Subsequently two transports, an old model B-17D and the emergency expedient of six A-24s were made available for transport duties at Port Moresby. As a precaution, further shielding revetments were built at Seven-Mile to protect transports and other visiting aircraft.

Meanwhile substantive reinforcements were destined for the beachhead area. A sixth convoy comprising three transports and escorts is listed below:

Sixth Buna Convoy

- *Tenryu* (4,350-ton light cruiser; built 1919; 4 x 5.5-inch guns; 6 x torpedoes)
- *CH-22, CH-23, CH-24* (438-ton submarine chasers; built 1941; 1 x 3-inch gun)
- W-20 (648-ton minesweeper; built 1941; 3 x 4.7-inch HA guns; 2 x 25mm AA guns)
- *Kazuura Maru* (6,804-ton IJA transport; built 1938)
- *Ryoyo Maru* (5,974-ton IJA Transport; built 1931)
- A third transport (sometimes misnamed in post-war histories as *Kanyo Maru*, but which is unidentified)

This sixth convoy departed Rabaul in the early hours of 17 August and arrived off Buna at dusk on the following day. It carried the main strength of the South Seas Force as well as additional base supplies for the Buna airfield. This convoy was not spotted by Allied aircraft and after unloading returned safely to Rabaul.

To further bolster this landed force now in Papua the main strength of the 41st Infantry Regiment (less one battalion) had arrived in Rabaul from the Philippines and was loaded on the seventh convoy as listed below:

Seventh Buna Convoy

- *Tsugaru* (4,400-ton mine-layer; built 1941; 4 x 5-inch guns; 4 x 25mm guns; 1 x E7K Alf floatplane; used as AA ship)
- *CH-28, CH-29, CH-30* (420-ton submarine chasers; built 1942; 1 x 3-inch gun)
- *Kiyokawa Maru* (6,860-ton seaplane tender; built 1937; 2 x 5.9-inch guns; used as transport with aircraft at shore bases)
- *Myoko Maru* (4,103-ton IJA transport; built 1939)

This seventh convoy made use of the spacious decks of the seaplane tender *Kiyokawa Maru* for transporting cargo as her air component was operating from scattered shore bases at this time. This convoy departed Rabaul on 19 August and arrived off Buna late on 21 August. Air cover was provided by two *shotai* of No. 2 *Ku* Zeros from Lae commanded by Warrant Officer Wajima Yoshio. Several ships were spotted by Allied reconnaissance off Buna on 21 August, but it was too late in the day for any attacks. The ships departed for Rabaul the following day and all arrived safely.

Meanwhile, the safe arrival of these latest Buna convoys was a significant achievement for the Japanese, but also reflected the fact that the Allies were struggling to sustain a high tempo of operations after their response to the initial Buna landing and then the many missions in support of American invasion of Guadalcanal. The personnel and supplies delivered by these convoys enabled the Buna airfield to become operational, with detachments of Model 21 and 32 Zeros from the No. 2 *Ku* and the Tainan *Ku* arriving there on 22 August. The Tainan *Ku* created its own separate Buna detachment from Model 32 Zeros, forming a separate command structure and markings regime different to those markings applied to its Lae and Rabaul-based Model 21s.

In the days leading to 22 August Allied offensive operations were very modest and consisted only of bombs dropped by B-17s on reconnaissance missions. One such B-17 dropped 4 x 500-pound bombs on Kavieng's airfield on 18 August, where increased activity had been noted.

Debris litters the Seven-Mile runway following the Japanese air raid on 17 August. Most of the transports in the background had to fly back to Australia for repairs after being hit by shrapnel. A single P-39 Airacobra passes overhead.

Limping Lizzie was among the second batch of F-4 reconnaissance Lightnings delivered to Melbourne on 12 August 1942. It is seen at Fourteen-Mile shortly afterwards while serving with the 8th PRS.

Meanwhile, the 8th Photo Reconnaissance Squadron received a major boost to its staffing complement when on 27 July it was reinforced with 152 men. These included several photographers from the 435th BS and about twenty additional pilots. The idea was that as the 8th PRS assumed more reconnaissance duties, the 435th could increasingly focus on bombing missions. Most of the new detachment arrived at Melbourne by ship, led by Lieutenant Paul Staller from Pennsylvania. An advance contingent of the fresh pilots was flown to Townsville the next day just in time to experience the series of Emily flying boat raids, described in Chapter 7.

Although at this stage the 8th PRS was exclusively operating F-4s (the photo reconnaissance version of the P-38 Lightning), a B-17 for use as a long-range platform was delivered to the unit on 31 July. This was a worn B-17E, serial 41-2458, which had been evacuated from Java and the unit's first job was to remove the lower turret and replace it with camera platforms. The first few days of August were spent overhauling the small and original fleet of F-4s which had seen much use over July, however a batch of new F-4s had arrived by ship in Melbourne but needed unloading and assembly at the nearby Commonwealth Aircraft Corporation factory.

Meanwhile, the new F-4 pilots commenced transition training at Townsville on the B-17 including Staller. Their new charge had no nickname however crews soon came to calling it

Old 58 (it would later serve as *Yankee Diddler* with the 43rd BG), and on 7 August it undertook its first assignment which was the carriage of photographic and dark room equipment from Townsville to Horn Island. The bomber wrestled getting back to Townsville in atrocious weather and diverted to Charters Towers as an alternate. This too proved too difficult, so they climbed above a thick overcast and finally made Garbutt. It was a rough introduction to the theatre's capricious weather for several "green" crewmembers.

Nonetheless, Staller was keen to get his first mission underway, and on 13 August took off in F-4 Lightning 41-2125 from Townsville at 1045. The Lightning had just had its engines overhauled, and he first staged to Horn Island where he arrived at 1700. The idea was to reach Port Moresby the next day where he was to be briefed on a reconnaissance mission to Lae. Staller could have overnighted at Horn Island, but his enthusiasm saw him proceed onwards to Port Moresby at dusk, a journey of two hours flight time. The flightpath was almost due east and for unknown reasons Staller failed to find his destination in the dark. He continued on the same track, and flew past Port Moresby for an hour or so. Staller realised by that stage that he must have missed Port Moresby, and instead found Misima Island, in the Louisiade Archipelago to the east of Milne Bay. He started circling the island where his F-4 was observed by an Australian observation detachment based on a coastal part of the island below. The western section of Misima Island has a mountain spine reaching to 2,500 feet, and Staller flew into one of these mountainous valleys in instrument conditions, followed by a crash. The Australian detachment was instructed to find the wreckage of Staller's F-4 which they finally located with the assistance of locals on 18 September. The F-4 had flown through trees and struck the ground hard, with the impact tearing off both engines. An injured Staller had clearly survived the crash, as his body was recovered some distance from the F-4's gondola, along with papers and maps from the cockpit. This was second time the 8th PRS had lost an F-4 on combat duty, following a loss during one of the unit's very first missions on 4 May.

Fortresses in freshly bulldozed revetments at the western end of Seven-Mile 'drome, a series of which were built in response to the losses of valuable transport aircraft suffered on 17 August.

Tainan Ku pilot FPO2c Kakimoto Enji prepares to ditch his Model 21 Zero V-130 in Milne Bay on 27 August 1942. Kakimoto observed correctly in his interrogation report that his fighter had already seen "a good deal of service".

CHAPTER 11

EMERGENCY!
MILNE BAY 23 AUGUST – 8 SEPTEMBER

As noted in Chapter 10, from mid-August there were strong indications of an imminent Japanese offensive in the New Guinea theatre, and accordingly the garrison at Milne Bay was put on high alert. Ground forces at Milne Bay had been strengthened with the arrival of the 18th Brigade of the Australian Imperial Force, part of the veteran 7th Division which had returned to Australia from the Middle East in recent months. By 22 August, Milne Force, commanded by Major-General Cyril Clowes, comprised some 8,824 troops. However only about half of this total was infantry, and it included 1,365 Americans, mostly from engineer units. Importantly, the Milne Force command included ground and air force units, so that Clowes had direct control of the Kittyhawks of Nos. 75 and 76 Squadrons and the forward detachment of Hudsons from Nos. 6 and 32 Squadrons.

The awaited Japanese offensive would indeed target Milne Bay, and was both two-pronged and an all-naval affair: no IJA troops would be utilised. This gave the IJN planners at Rabaul (and Tokyo) flexibility to quickly enact a plan without lengthy and formal negotiations with their IJA counterparts. The intention had originally been to establish a forward seaplane base and an 800 metre long airfield on the small island of Samarai, but this was changed to the occupation of nearby Milne Bay following the discovery of the Allied base there in early August. Because Milne Bay had only been established for a short time, it was assumed that ground defences there were relatively thin and hence the limited number of naval troops available would suffice for the task.

The capture of Milne Bay airfield also offered another irresistible strategic objective: it would provide a forward airfield to facilitate the capture of Port Moresby, an objective still exercising the minds of Japanese high command. The destruction of the Allied cruiser squadron off Savo Island on the night of 8/9 August also gave Vice-Admiral Mikawa the encouragement he needed to launch the campaign. Mikawa assessed that without cruisers, the USN was unlikely to send its carriers to resist the Milne Bay invasion.

The first part of the IJN plan was innovative, using barges to transport part of the 5th Sasebo SNLF (353 men) from Buna to a landing point on the Papuan coast. These troops could then advance a short distance overland to threaten Milne Bay from the north, with the barge movement including an overnight stop at Goodenough Island. However, the main operation would be via a conventional convoy that would land a larger force of troops on the coast just east of the Milne Bay base. This comprised 612 men of the 3rd Kure SNLF and 197 men from the 5th Sasebo SNLF, together with 362 construction troops from the 10th Establishment Unit. The units would take with them two light tanks, two 37mm guns and two 70mm artillery pieces. These movements were timed to coincide with the arrival of Zeros at Buna, where detachments from the Tainan

RAAF Kittyhawks parked in dispersals at Milne Bay in September 1942. During the peak of the Milne Bay fighting they were evacuated nightly to Port Moresby. Note they have not yet lost their red roundels.

and the No. 2 *Kokutai* had arrived on 22 August, as noted in Chapter 11.

The day after their arrival at Buna, on 23 August, eight No. 2 *Ku* Zeros combined with an identical number of Tainan *Ku* fighters for a sweep over Milne Bay. Despite searching for 35 minutes, the usual cloud cover frustrated their goal, and the incursion was not even noticed by those on the ground. However, the Zeros did briefly skirmish with a lone No. 75 Squadron Kittyhawk returning from a patrol, flown by Flight Lieutenant Frank Coker. When the Zeros returned to base, the somewhat rudimentary runway at Buna claimed a victim when the Zero flown by FPO2c Yamazaki Ichirobei was badly damaged in a landing accident, and the fighter was written off.

The following day, 24 August, saw the barge movement from Buna commence and a raid against Milne Bay was attempted, using Bettys from Rabaul combining with Zeros from Buna. After leaving Vunakanau, at about 0800 that morning seven Bettys from the Kisarazu *Ku* combined with seven from the No. 4 *Ku* and headed for Milne Bay. The Kisarazu *Ku* aircraft were from the No. 2 *chutai* under Lieutenant Nakamura Tomo'o which had had only flown in from the Marianas the previous day. The importance of the raid to the Japanese is underlined by the fact that the Bettys were led by the Kisarazu *Ku Hikotaicho* Lieutenant-Commander Watanabe Hatsuhiko. However, they soon encountered atrocious weather and after failing to get around it all of the Bettys had returned to Vunakanau before midday.

However, the Zeros persisted and found a gap through the weather near Milne Bay. This was a combination of fifteen Tainan and No. 2 *Kokutai* fighters which tangled with fifteen No. 75 Squadron Kittyhawks and eleven from No. 76 Squadron which were airborne on patrol. In cloudy conditions some heated combats took place and the Zeros expended over 5,000 rounds of ordnance. However, despite claims from both sides the only result was minor damage to a P-40 and a Zero. The confused encounter was summarised by No. 76 Squadron's Flight Sergeant Sullivan:

> Number 2 from this pair got onto my tail and fired from 100 yards. Lousy shot and would not follow into cloud. Came out of cloud and stooged towards ranges where odd brawls were taking place. Fired on by Kittyhawk.

The Allies received the first concrete evidence of the Japanese offensive operation when late in the afternoon of 24 August a coastwatcher reported Japanese barges moving down the northern coast of Papua. However, no air attack could be launched because of the evening fading light and also because the Kittyhawks at Milne Bay had been busy with the Zero combat. That night the barges crossed over to Goodenough Island, arriving at dawn the following morning. However, there was nowhere to conceal the barges, which were simply pulled up onto a beach. This was the first multi-day mass barge movement in the New Guinea theatre, and in due course the Japanese would become more adept at such operations. However, leaving the barges unconcealed in this manner would lead to disaster.

On the morning of 25 August Milne Force received a message from a coastwatcher on Goodenough Island about the presence of the seven barges. Nine Kittyhawks from No. 75 Squadron were sent to the island in two flights under the respective leadership of Flying Officers Piper and Atherton. The barges were found easily and were seen to be packed with equipment although few troops were visible. While one flight acted as top cover, the other strafed the barges from low level at 1030. After a total of six runs the barges were left blazing.

In fact, all of the barges had been destroyed along with many of the provisions including the sole radio transmitter. Hence 353 men of the 5th Sasebo SNLF, known as the Tsukioka Unit after their CO Commander Tsukioka Torashige, were effectively marooned on Goodenough Island. Worse, headquarters in Rabaul had no idea of their whereabouts and as the Milne Bay battle evolved it was falsely believed that these troops had landed as planned on the Papuan coast to the north of Milne Bay and were operating under conditions of radio silence. Only when Tsukioka Unit messengers succeeded in reaching Buna by native canoe on 9 September were arrangements made to collect these marooned marines. However, those operations became a long and costly saga beyond the parameters of this volume, and in the end the marines were not rescued until late October.

Meanwhile, the main Milne Bay invasion force had departed Rabaul at 0700 on 24 August, with the convoy comprising the ships listed below:

Milne Bay Invasion Convoy

- *Tenryu* (4,350-ton light cruiser; built 1919; 4 x 5.5-inch guns; 6 x torpedoes)
- *Tatsuta* (4,350-ton light cruiser; built 1919; 4 x 5.5-inch guns; 6 x torpedoes)
- *Hamakaze, Tanikaze, Urakaze* (2,033-ton destroyers; built 1941; 6 x 5-inch DP guns; 4 x 25mm AA guns; 8 x torpedoes)
- *CH-22, CH-24* (438-ton submarine chasers; built 1941; 1 x 3-inch gun)
- *Kinai Maru* (5,040-ton IJN auxiliary transport; built 1930)
- *Nankai Maru* (5,114-ton IJN auxiliary transport; built 1933)

The 1,171 men of the invasion force were embarked on the transports *Kinai Maru* and *Nankai Maru*. Escort was built around the two elderly light cruisers *Tenryu* and *Tatsuta*, as well as three modern destroyers *Hamakaze, Tanikaze* and *Urakaze*. These destroyers were the first modern examples used in New Guinea convoys and boasted a respectable AA armament of six 5-inch Dual Purpose and four 25mm AA guns.

At 0830 on 25 August the convoy was spotted by Allied reconnaissance aircraft near Kitava Island, some 100 miles north-east of Milne Bay. Subsequent contact with the convoy made it clear to Milne Force headquarters that the destination was Milne Bay that night. Initial attempts to launch air attacks on the convoy were frustrated by bad weather, but a mission by half a dozen No. 75 Squadron Kittyhawks led by a Hudson found the convoy and attacked it at 1507. The Kittyhawks were flown as fighter-bombers, with the mission described by Pilot Officer Jeff Wilkinson:

> We were all bombed-up with 250-lb bombs but unfortunately none of us had ever dropped a bomb from a Kittyhawk before. We were told that the only way to do it was to line up the target – go into a steep dive – start pulling out and count three and pull the lever and let the bomb go. We went out with FO Robertson and his lone Hudson – he was to lead us and find where the Japanese convoy was. After about half an hour of flying through very bad weather we came right over the Japs. The two big cruisers were making half circles and firing lots of flak at us. There seemed to be numerous small ships around and at least two transports. We went in one after another and I went down firing all my guns at a transport that was heavily laden with Japanese. Dropped my bomb and pulled around.

A second Kittyhawk attack by nine machines from Nos. 75 and 76 Squadrons followed an hour and a half later at 1642. Then in conditions of driving rain, failing light and with a low ceiling of just a few hundred feet, Hudson A16-205 flown by Pilot Officer Law was one of two sent to make at masthead attack at 1800. The convoy was located approaching the entrance to Milne Bay, and a mix of 100- and 250-pound bombs were dropped from just 100-feet, during which a "formidable barrage" of AA of all calibres was experienced.

Subsequently Law found that the Milne Bay airfield was closed, and he and his crew only managed to find Port Moresby in the darkness with the aid of Radio Direction Finding. While the crew was being debriefed after landing at Seven-Mile:

> … the US refuelling tanker pulled up as usual … after a few minutes one of the tanker crew loudly requested the capacity of the Hudson's front port tank. Someone told him 120 gallons, to which he replied that he had pumped 450 gallons already and the tank was not full. By this time the plane was standing in a pool of 100 octane petrol and an inspection disclosed a hole under the tank through which you could pass your arm to the armpit.

After flying to Horn Island the Hudson was directed to Townsville for repairs where 102 hits were counted in the airframe. While no telling blows were landed on the convoy during these attacks, some 20 marines were killed. The destroyer *Urakaze* recorded minor damage from strafing and suffered one dead and three wounded. During the afternoon nine Fortresses from the 93rd BS at Mareeba had also set out to attack the ships but failed to find the convoy amid the bad weather.

On approach of the convoy, the sole Allied naval presence in Milne Bay, the destroyer HMAS *Arunta*, was ordered to escort a merchant ship back to Port Moresby. This left an RAAF launch as the only means of observing the entrance to Milne Bay and shortly before midnight the crew reported the arrival of Japanese ships. As the Japanese troops went ashore in the early hours of 26 August, the warships bombarded what they believed was the airfield. However, partly due to heavy fog, the Japanese were unable to accurately fix their position and they landed several miles further east than intended.

Nine 93rd BS Fortresses departed Mareeba at 0510 hours, but after one abort back to Mareeba in the bad weather, eight arrived individually between 0630 and 0745 over Milne Bay. They made individual attacks on ships under a cloud base as low as 2,000-feet. Attacking at such low level the bombers encountered heavy AA fire, and several were damaged and suffered casualties. A relatively new "F" model bomber, 41-24354, flown by Captain Clyde Webb, took a direct hit in the No. 2 engine and then drove almost vertically into the ocean, leaving behind a burning gasoline slick. All aboard were killed, giving the aircraft the dubious distinction of being the first B-17F lost in the Pacific.

A B-17E, serial 41-2621, named *The Daylight Ltd* commanded by Captain Ken Caspar dropped bombs on a cruiser and then a transport from around 200 feet, but with no results apparent. Caspar's Fortress sustained enemy fire from the ships, receiving numerous hits which wounded several crew including a side-gunner wounded in the eye. The gunfire also disabled the hydraulic system and Caspar figured the medical facilities were better at Mareeba than Port Moresby so

B-17E The Daylight Ltd as it came to rest at Mareeba on 26 August 1942 after being damaged from AA fire while attacking enemy ships at Milne Bay.

he elected to return directly to the North Queensland base. With no flaps he was forced to make a "hot'" landing at around 130 miles per hour. The starboard main landing gear collapsed, skewing the bomber to the right in an uncontrolled ground loop. It slammed into gum trees adjacent to the runway, becoming a total write off.

Nonetheless, the early appearance of the B-17s convinced the Japanese to abandon landing operations and withdraw their ships, although most of the personnel and both Type 95 light tanks had got ashore. The *Nankai Maru* received some light damage, and both it and the *Kinai Maru* sailed back to Rabaul escorted by the two submarine chasers *CH-22* and *CH-24*. The two light cruisers and three destroyers remained at sea that day but in the general vicinity of Milne Bay. Attempts to find the ships after the initial B-17 attacks failed, as they were screened by thick cloud and fog.

At first light the sole Hudson still at Milne Bay, A16-185, had made a reconnaissance which found the landing site crowded with fifteen barges moored inshore. Scattered stores could be seen piled up, including a large number of fuel drums, some of which were floating next to the barges. The troops had landed on a narrow coastal strip lying between steep jungle-covered mountains and the shore. There was some sporadic fighting between forward patrols of both sides, although all troops were seriously impeded by the sodden conditions and the Japanese made their main movements at night.

That morning the Kittyhawks made repeated attacks on the landing site, in conditions later described by an Australian general as an "airman's paradise". Within minutes of taking off, the pilots were over the target area which was easily identified. Machine gun fire and bombs ripped into the barges which were soon all sunk, and large fires began as the fuel drums were hit. Many of the pilots made multiple attacks, with No. 75 Squadron alone flying 26 sorties. The Kittyhawks were joined by a No. 6 Squadron Hudson from Port Moresby which also bombed the landing area.

No Japanese aircraft made it to Milne Bay on 26 August, largely due to a successful Airacobra strafing attack on the airfield at Buna that morning as will be described in Chapter 13. The only loss suffered by the RAAF in these attacks was the Kittyhawk A29-110 ("J") flown by Flying Officer Alan Whetters, who ran short of fuel trying to return home in the poor visibility. Eventually he ditched near a jetty on Sideia Island off Milne Bay's eastern-most cape. Whetters was rescued by natives in a dinghy, who took him to the nearby Catholic Mission. Assisted by missionaries there, Whetters built a log raft in an attempt to float his Kittyhawk back to base, however when he realised the proximity of Japanese forces he scuttled the raft. He returned to his unit later that afternoon.

Meanwhile 22nd BG Headquarters at Townsville had been warned of the invasion and sent seven B-26s from the 2nd and 19th BS to Port Moresby on the afternoon of 25 August. The next morning these seven headed for the Japanese shipping but instead encountered severe thunderstorms about fifty miles from the target, with cloud base down to sea level. All had returned to Seven-Mile by noon. After refuelling, three of these B-26s led by Major Walter Greer headed for Mullins Bay, about a dozen miles from the RAAF airfield, where a 10,000-

ton transport had been reported. This time they got through and swept the area from about 2,000 feet. Instead of the transport they found only the Japanese barges on which they dropped their bombs. That same afternoon Townsville mustered every B-26 it could, and a dozen made the journey to Port Moresby that afternoon of 26 August. Meanwhile, a second B-17 strike by nine aircraft from the 28th BS failed to find any of the ships and instead they too unloaded their bombs over the landing area.

That night the two light cruisers and three destroyers returned to Milne Bay but were unable to make contact with the Japanese forces on land. The ships bombarded what they thought was the airfield, with the shells instead landing near the forward Australian positions around KB Mission. The shelling caused no casualties, although the Australians were forced back by Japanese ground forces that night. By dawn the ships had withdrawn once again.

The morning of 27 August brought the fourth Japanese air raid on Milne Bay. This consisted of eight No. 2 *Ku* Vals, each armed with two 60-kilogram bombs, escorted by a seven Zeros from both the Tainan and No. 2 *Kokutai*. The three Val *shotaicho* were experienced fliers who had all survived the disastrous one-way 7 August Guadalcanal mission which ended in ditching at the Shortlands seaplane base. However, most of the other Vals were manned by relatively inexperienced replacement crews. The Vals arrived over the target area before the Zeros and completed their attacks at around 0815. Some bombs landed astride the No. 1 Strip runway but did no damage.

Several of the Zeros then arrived and their attention was drawn to the crippled LB-30 Liberator AL515 *Yard Bird*, which was strafed and set on fire in the face of heavy small arms fire from the ground. Meanwhile a single Zero had spotted a lone B-17F from the 435th BS on a reconnaissance mission. Despite the Fortress pilot quickly pulling up into cloud the Zero managed a single pass which caused some minor damage to the bomber. One of the crew

A Japanese barge and stores lie abandoned at Milne Bay after being wrecked during a devastating series of attacks by RAAF Kittyhawks on 26 August. The photo was taken shortly after the area was reoccupied by Allied forces in early September.

members got a clear photo of the attacker, showing a double chevron on the rear fuselage. This was the Model 21 fighter of *chutaicho* Lieutenant Yamashita Joji, flying *Hokoku* fighter HK-529.

After strafing the airfield, two of Yamashita's Zeros were drawn to a flight of Marauders which were searching for enemy shipping in the area. The Marauders were from a dozen which had departed Port Moresby but had since split into two formations and were searching at low altitude. However, both Zeros had been hit by ground fire including Lieutenant Yamashita Joji mentioned above whose Zero trailed fuel and soon crashed despite a last-ditch attempt to attack a B-26.

The other Zero was flown by FPO2c Kakimoto Enji and when chasing the B-26s Kakimoto realised he was losing oil pressure and his engine would soon shut down. Kakimoto successfully ditched his Zero (tail-code V-130) and swam ashore, confident he would meet friendly forces. After spending a few days in a native village, Kakimoto was turned over to Australian authorities and was subsequently sent to the POW camp at Cowra, New South Wales. Kakimoto was a ringleader during the infamous 1944 Cowra Breakout when he hung himself.

Meanwhile arriving to a scene of scattered Vals and Zeros over Milne Bay was a patrol of six No. 75 Squadron Kittyhawks led by their CO, Squadron Leader Les Jackson. Jackson and his wingman, Sergeant Roy Riddel, targeted two Zeros that were busy trying to strafe Kakimoto's ditched Zero in order to prevent it falling into Allied hands. After closing to close range, the surprised Japanese made easy targets for the Australians and both FPO1c Yamashita Sadao and Flyer1c Ninomiya were killed.

Other Kittyhawks targeted the Vals, and that crewed by pilot FPO2c Takahashi Koji and observer Lieutenant Yoshinaga Hiroshi was downed with the loss of both men. A second Val was forced to ditch to the east of Milne Bay, but the pilot Flyer2c Shibuya Masakichi did not survive. His observer, Flyer1c Koyamada Masami managed to reach the shore but was subsequently captured by Australian troops. Koyamada served out the war in captivity before returning to Japan in 1946.

Overall, the encounter had been a great victory for the No. 75 Squadron pilots, with the Japanese losing two Zeros and two Vals to the Kittyhawks. Added to these losses were the other two Zeros downed due to ground fire. However sadly for the Australians they lost two Kittyhawk pilots on this same day, one of whom was arguably their most senior and capable pilot.

There were no witnesses to the loss of the first Australian, Pilot Officer Stuart Munro flying A29-108 *Schuftie*. It is believed he was downed by a Zero, and his crash site was found in the jungle two months later. The second pilot was lost in an incident unrelated to the air combat. This was No. 76 Squadron's CO and ace Peter Turnbull who was flying low over enemy positions searching for tanks when he crashed into trees and was killed, probably after being hit by ground fire.

Subsequently the Milne Bay No. 3 Strip was named Turnbull Field in his honour. No. 1 Strip was named after another RAAF squadron CO, that of No. 33 Squadron, Charles Gurney, who had been killed while flying as the co-pilot of a B-26 on 2 May.

LB-30 Liberator AL515 Yard Bird burns after being strafed by Zeros on No. 1 Strip, Milne Bay, 27 August 1942.

Below: Model 21 Zero tail-code V-130 which was ditched by FPO2c Kakimoto Enji in Milne Bay on 27 August. It was raised several weeks later by Australian soldiers.

A6M Model 21 Zero V-117, CN 2641, Tainan Ku chutaicho Lieutenant Yamashita Joji shot down Milne Bay 27 August 1942.

Further Kittyhawk and Hudson missions against suspected ground positions were flown during the day, although there were few landmarks to assist targeting and on occasion friendly positions were strafed. Meanwhile the formation of cruisers and destroyers sailed back to Rabaul, but only after detaching the destroyer *Hamakaze* to return to Milne Bay that night. The destroyer tried to re-establish contact with the ground forces, which had been sporadic due to radio failures. However, the use of signalling lights also failed due to heavy rain, and the vessel departed for Rabaul at 0230 on 28 August.

This was the third night in a row that Japanese ships had appeared in Milne Bay, and these appearances had the Australians believing that the ground forces were being reinforced and resupplied. Meanwhile, at Rabaul Vice-Admiral Mikawa had been fully expecting his ground forces to have captured Milne Bay airfield within 24 hours of landing. It was not until the afternoon of 27 August that Mikawa learned that the attack had been unsuccessful, and his staff made urgent arrangements to send reinforcements.

By dawn on 28 August the Australian ground forces had withdrawn to a perimeter close to the boundaries of No. 3 Strip. Clowes was reluctant to aggressively commit all of his forces to battle, including his best AIF battalions, as he feared another landing and the saturated conditions made large scale movements difficult in any case. However, the pressure was being felt throughout the garrison and especially by the RAAF ground crews which were working tirelessly to refuel and rearm aircraft, including the never-ending task of re-belting 0.50-inch ammunition. All of this was done in wet and steamy conditions where little sleep was possible, and malaria was a constant threat.

As a precaution against a breakthrough to No. 1 Strip, on the afternoon of 28 August, 30 Nos. 75 and 76 Squadron Kittyhawks were flown to Port Moresby. However dark low cloud greeted the incoming pilots at Seven-Mile shortly before dusk, and Sergeant William Cowe was killed when he flew A29-109 into a hill, in the circuit area of Seven-Mile, just to the south of the field. At the same time a Hudson evacuated fifteen pilots from Milne Bay to Port Moresby, as described by Pilot Officer AJ Gould:

> We were heavily overloaded, and I lay face down in the bomb-aimer's position with one or two chaps on top of me. On take-off I watched the coconut trees at the end of the strip loom closer and we actually caught the tops of them on staggering into the air. A Kittyhawk flown by Bill Cowe of 75 formated on us to ensure he could find Moresby airfield in the dark. He broke away over the strip but flew into a surrounding hill and was killed.

The Kittyhawks would return to Milne Bay the following morning, and the process would be repeated each day. During one of these return flights on 30 August Squadron Leader Les Jackson force-landed Kittyhawk A29-71 "B" on the coast east of Port Moresby after experiencing mechanical problems. He was subsequently returned to Port Moresby by a lugger after a short stay in a native village, and his fighter was later recovered and repaired by a RAAF salvage unit.

On this same afternoon of 28 August that the Kittyhawks first evacuated to Port Moresby, Mikawa's Milne Bay reinforcement convoy departed Rabaul, comprising the following ships:

Milne Bay Reinforcement Convoy

- *Tenryu* (4,350-ton light cruiser; built 1919; 4 x 5.5-inch guns; 6 x torpedoes)
- *Arashi, Tanikaze, Urakaze* (2,033-ton destroyers; built 1941; 6 x 5-inch DP guns; 4 x 25mm AA guns; 8 x torpedoes)
- *Yayoi* (1,772-ton destroyer; built 1926; 4 x 4.7-inch guns; 6 x torpedoes)
- *Murakumo* (2,090-ton destroyer; built 1929; 6 x 5-inch DP guns; 9 x torpedoes)
- *PB-36, PB-38, PB-39* (935-ton patrol boats; built 1921-22; 2 x 4.7-inch guns)

Unlike the earlier convoy, this one included no transports with the reinforcements instead loaded aboard destroyers and patrol vessels. The reinforcements comprised 567 men from the 3rd Kure SNLF and 200 men from the 5th Yokosuka SNLF. The patrol vessels were former destroyers, capable of 18 knots, so that the entire convoy was relatively fast.

On the afternoon of 29 August a No. 6 Squadron Hudson, flown by David Colquhoun, made a reconnaissance flight to St George's Channel, just short of Rabaul itself. On the return flight at 1550 Colquhoun came upon the reinforcement convoy which was accurately assessed as one cruiser and eight destroyers. A bombing run was made against the second last destroyer from 3,000 feet, with four 250-bombs believed to have landed near the ship's stern. However, no damage was inflicted.

After landing at Milne Bay, the Hudson was loaded with bombs and Colquhoun was instructed to lead six Kittyhawks to jointly dive-bomb the convoy. However, amid squalls and near darkness nothing was seen and the aircraft returned to base. Unfortunately, during this same evening, Pilot Officer Brendon Davis landed on the wrong side of the flare path and was fatally injured when Kittyhawk A29-106 collided with a tree. Davis, a ferry pilot, was delivering the aircraft back to the combat zone after repairs at Port Moresby.

That evening the Japanese reinforcements and supplies were unloaded before midnight. The shore area was again bombarded but the shells landed harmlessly among the hills behind the base area. Not so lucky was the crew of an RAAF launch who were evacuating to a safer harbour outside of Milne Bay. Near the entrance to the bay the launch was caught in a searchlight of a Japanese warship and suffered a direct hit from the fourth shell fired at it, killing three of the five onboard.

The following day Fortresses from the 28th and 30th Bombardment Squadrons attempted to find the Japanese ships but were again frustrated by the weather. Otherwise, Kittyhawk ground attacks continued as usual. The Japanese ground forces had dispersed themselves deep in the jungle in preparation for an offensive push against the main defensive perimeter that night. The main attack came at 0300 on 31 August, and flares fired by the Australians revealed groups of attackers advancing from the eastern edge of No. 3 Strip. However, the Australians were able to respond with heavy weapons. Artillery, mortar and machine gun fire took a heavy toll on the Japanese, and the attack was thwarted.

A 1943 view of No. 3 Strip, Milne Bay (Turnbull Field). Japanese forces landed on the coast visible on the left and pushed right up to the edge of the strip in the early hours of 31 August. They were then forced to withdraw after running into Allied fixed defence positions equipped with heavy weapons.

At 0800 on 31 August Commander Yano Minoru of the 3rd Kure SNLF reported:

> We attacked as planned but came under heavy crossfire from well dug-in positions when we drew near the enemy, sustaining heavy casualties. Though the reserve units were summoned, they did not arrive before dawn, making the assault problematic.

That morning the Japanese had planned to counter Allied air attacks with an offensive sweep by seven Tainan *Ku* Zeros. These arrived over Milne Bay at 1140 and were engaged by AA batteries, but took avoiding action. The Zeros could not see any aircraft present so didn't strafe. Indeed, the Kittyhawk force was just then flying back from Port Moresby.

After the failure of the Japanese attack, Australian forces began pushing steadily eastwards. However, among other things the Japanese falsely believed the Tsukioka Unit was approaching overland from the north at this time when in fact it was marooned on Goodenough Island. There was also hope that IJA troops might be soon available to help reinforce and rectify the situation, together with planned air and naval support. In the meantime, existing forces would fall back to a holding position in the jungle east of the Milne Bay base.

At dawn on 1 September six Zeros and five Vals, all from the No. 2 *Ku*, departed Rabaul for Milne Bay but were forced to return to base in the face of bad weather. On this same day Mikawa decided to send 130 men from the 5th Yokosuka SNLF to Milne Bay as reinforcements, and these departed Rabaul on the morning of 2 September aboard the patrol vessel *PB-39* escorted by several destroyers. Also on 1 September seven No. 75 Squadron Kittyhawks led by Squadron Leader Jackson made attacks on what was believed to be a Japanese headquarters location, but the jungle was so dense that results were unclear.

However, at this same time came news that an Allied transport and a "cruiser" had entered Milne Bay. This was the merchant ship *Tasman*, desperately needed at Milne Bay where the garrison had less than 20 days rations remaining. It was escorted by the destroyer HMAS *Arunta*. On receipt of this news Rabaul quickly collected one of the most eclectic collections of IJN aircraft yet assembled in the South Pacific to strike the ships. The force comprised six No. 2 *Ku* Vals, five No. 4 *Ku* Bettys, one Toko *Ku* Emily flying boat, and eight Zeros from both the No. 2 and Tainan *Kokutai*.

The mixed collection of aircraft departed Rabaul at various stages in mid-afternoon, but soon ran into bad weather. The five Bettys were led by the *Hikocho* of No. 4 *Ku*, Captain Moritama Yoshimitsu. Despite the unit being known as the "Moritama *Kokutai*" it was extremely rare for Moritama to take to the air, but his presence underlined the importance that Rabaul's headquarters placed on salvaging the dire Milne Bay situation. However, Moritama's Bettys turned back after encountering a front at 1625, only one hour after departing Vunakanau.

Moritama was not the only senior IJN officer airborne that afternoon. The *Hikotaicho* of the Toko *Ku*, Commander Onaka (first name unknown) was aboard the sole Emily with a crew of nine which left Rabaul at 1530. Although encountering the adverse weather system two hours later, Onaka persisted unsuccessfully looking for the reported shipping well into the evening and did not return to Rabaul until 2110.

In the end, all but a *shotai* of three Vals returned to base. Under the leadership of Warrant Officer O'Ota Gengo, who had been impressed with the critical nature and importance of the target, his Val *shotai* pressed on and made a determined search for the ships. However, after visiting Milne Bay that morning, both the *Tasman* and *Arunta* had since sailed out of the bay.

Having failed to find any ships and running low on fuel, in fading light just before dusk O'Ota led the three Vals down to land on a beach on the south coast of Papua. After removing survival packs from the dive-bombers, the crews tried to set them on fire before starting to trek overland to Buna. The Vals were soon spotted by a Kittyhawk on a ferry flight from Port Moresby to Milne Bay, and then visited by RAAF officers which landed a Tiger Moth alongside the dive-bombers. Although some parts of the Vals were damaged by fire, the remnants were later shipped to Brisbane for intelligence analysis. A few days after landing O'ota and his men were killed in shootouts with Australian soldiers.

Meanwhile the landing of the reinforcements from the *PB-39* was cancelled late on 2 September, as the Japanese made rapid changes to their plans in the face of an agreed greater priority ongoing at Guadalcanal. Also on this day Lieutenant General Sydney Rowell, who had only recently assumed command of the Australian Army in New Guinea, arrived in Milne Bay via a Hudson. His account of the flight is as follows:

> We flew down in a RAAF Hudson, and as we were approaching Milne Bay we met a tropical front. For three hours we flew at heights from 500 to 15,000 feet, looking for a break in the clouds. We were alternately bathed in perspiration and frozen. I told the pilot to go back to Port Moresby, but just then he found a break over China Strait and we flew in at sea level. As we landed on the pierced steel matting of the aerodrome we flew up a cloud of dust. I was there five hours, during which time about four inches of rain fell. As we took off we threw up, not dust, but a sea of liquid mud.

The following day, 3 September, saw the loss of No. 6 Squadron Hudson A16-220 which crashed into the ocean five miles from Port Moresby while on its way to patrol near Milne Bay. Pilot Officer William Campbell and his three crewmen were killed. Their bodies were recovered but the cause of the accident remains unknown, although it is likely that Campbell became disorientated in instrument conditions.

That night the destroyers *Arashi* and *Hamakaze* entered Milne Bay and embarked casualties and received first-hand information from the force ashore. The news was not good. Combat-fit Warrant Officers and higher ranks were down to one third strength, and of 560 combat troops remaining only about 200 were capable of battle. The destroyers bombarded the shore but once again without inflicting any damage or casualties.

On 4 September another attempted raid on Milne Bay by six No. 4 *Ku* Bettys escorted by nine Zeros was forced to turn back due to the weather. Once again, Rabaul had decided to have the mission led by a senior officer, with Kisazaru *Ku hikotaicho* Lieutenant-Commander Watanabe Hatsuhiko sitting in the lead No. 4 *Ku* Betty. The accompanying Zeros, from No. 6 *Ku*, were new players in the campaign, led by Lieutenant Tagami Takenoshin. Both units were back at Rabaul after encountering bad weather around midday and being airborne for two and a half hours.

A "cruiser" (possibly the destroyer *Yayoi*) was spotted by a Hudson from Port Moresby on the morning of 4 September and a strike by another three Hudsons was organised. However, one returned to base due to engine trouble and another ran low on fuel before spotting any ships. The third Hudson sighted the "cruiser" at 1310 and attacked it in conditions of low visibility but did not claim any hits.

Subsequently two late afternoon strikes were launched against the "cruiser" by six 3rd BG B-25s and five Hudsons. However, none of the aircraft made contact with the ships before nightfall, when downwards visibility became impossible. One Hudson then made fleeting contact by seeing a burst of AA fire below, but no attack followed.

In a disastrous return flight to Port Moresby, two 13th BS B-25Cs were lost, named *The Queen* and *Hell Cat*, and flown respectively by Captain Gustave Heiss and Lieutenant Hubert Rapp. The six Mitchells had headed for home in darkness with their bombs still aboard, hugging the coast while climbing for altitude. About halfway back Heiss turned on his landing lights and peeled away to starboard. Then he descended and ditched in the water below. No radio call was made by Heiss and no explanation survives as to why he broke away. Heiss and his crew, including RAAF co-pilot Flying Officer Allan Page, were never found. Meanwhile, low on fuel and in bad visibility, Rapp also broke away and ditched *Hell Cat* in shallow water off Kupiano village about halfway back to Port Moresby. *Hell Cat* broke up badly on impact killing all five crew. The next day, villagers recovered all their remains and delivered them to an ANGAU officer.

That night the destroyer *Yayoi* entered Milne Bay and embarked 224 casualties. Reports from the force ashore indicated that the situation had further deteriorated. By the following morning Rabaul took into account the parlous circumstances of the landing forces and made the momentous decision to evacuate in entirety. During the day of 5 September, the light cruiser *Tenryu* and the patrol vessels *PB-36*, *PB-38* and *PB-39* steamed towards Milne Bay but were not detected by Allied reconnaissance. That day Kittyhawks supported Australian troops advancing along the coastal strip towards the original Japanese landing point. However, that night the evacuation was carried out successfully. Some 1,318 men were evacuated on the

The disassembled engine from the No. 2 Ku D3A1 Val force-landed by Warrant Officer O'Ota Gengo on 1 September on a Papuan beach in Allied territory. The engine and other parts of the dive-bomber were transported to Port Moresby by lugger for examination by Allied intelligence.

Aichi D3A1 Val, Q-219, CN 3114, No.2 Ku, pilot FPO1c Maruyama Takeshi, force-landed on 1 September.

Tenryu and the three patrol vessels. In total the Japanese had lost some 625 men at Milne Bay.

With clear signs that the Japanese had been withdrawing, on the morning of 6 September the merchant ship *Anshun* was escorted into Milne Bay by the destroyer *Arunta*. *Arunta* left the bay during the day to cover two other ships waiting to enter the bay, but the *Anshun* stayed at the jetty unloading. However, there were indications from intercepted signals of continued Japanese naval activity in the area and three Hudsons were sent out to search the

A No. 6 Squadron Hudson takes off from a typically muddy Gurney Field, circa September.

approaches to Milne Bay. At 1535 A16-154 flown by David Colquhoun sighted a cruiser and a destroyer but a faulty radio meant the crew could not immediately report the find. Colquhoun continued to shadow the vessels for an hour in the face of AA fire, but by keeping his distance he avoided damage. After landing three Hudsons were sent out to attack the ships but failed to find them in conditions of poor visibility.

During the day six torpedo equipped Beauforts of No. 100 Squadron had arrived at Milne Bay from the mainland giving a significant boost to the RAAF's anti-shipping ability. These were accompanied by three Beaufighters from No. 30 Squadron, the very first of these twin-engine fighters to arrive in New Guinea.

However, these arrivals were too late to affect the arrival in Milne Bay that night of the light cruiser *Tatsuta* and the destroyer *Arashi* which had been spotted by Colquhoun earlier. Using searchlights, the ships shelled the *Anshun*, which rolled over and sank in shallow water with the loss of two of her crew. The hospital ship *Manunda*, anchored nearby and brightly marked with illuminated red crosses, was not targeted.

At first light on the next morning of 7 September Hudsons were out searching for the ships in atrocious weather, while a strike force comprising Kittyhawks, Beauforts and Beaufighters was kept ready at Milne Bay. Contact was made and then lost with the enemy ships that morning. They were located again in the afternoon and the strike force was launched. However, of the three Beaufighters launched, A19-13 commenced its take off run on an angle and slid off the metal runway before colliding with a parked Hudson. The accident wrote off the Beaufighter.

As eight Kittyhawks flew top cover, the two remaining Beaufighters and eight other Kittyhawks made strafing attacks on the two ships. The six Beauforts then made their attack runs, dropping their torpedoes from distances of between 1,200 and 1,700 yards. No hits were made with a report later blaming this on, among other factors, a lack of torpedo sights fitted to the Beauforts and the fact that the torpedoes were so ingrained with mud at Milne Bay that the depth setting mechanism could not be adjusted. Three Hudsons subsequently each dropped four 250-pound bombs on the ships but also without result.

That night two Japanese vessels again re-entered Milne Bay, probably the same cruiser and destroyer although their identities are unconfirmed. The vessels briefly shelled the base area once again before departing, wounding three men. It was the last incursion by Japanese vessels into the bay. The following day, 8 September, Hudsons searched for these vessels and amid continuing bad weather they were briefly found and bombed, but again without result.

The parting Japanese shot of the Milne Bay campaign was an air raid conducted on 8 September by nine No. 4 *Ku* Bettys escorted by nine No. 2 *Ku* Model 32 Zeros, all of which departed Vunakanau at 0935. Once again seconded to the No. 4 *Ku*, Kisarazu *Ku Hikotaicho* Lieutenant-Commander Watanabe Hatsuhiko commanded the mission which dropped six 250-kilogram and 71 x 60-kilogram bombs over Gurney Field around midday. These bombs landed along the eastern edge of the strip, destroying a slaughterhouse, two trucks and a small fuel dump. Three Americans, two AIF soldiers and two RAAF airmen were killed, along with several others who were wounded.

Lieutenant Kurakane Yoshio led the Zero escorts, which did not see combat but due to their short range returned via Lae to refuel while the Bettys returned directly to Rabaul.

The Anshun lies on her side at Milne Bay after having been shelled by the light cruiser Tatsuta and the destroyer Arashi on the night of 6 September.

Beaufighter A19-13 lies askew at Gurney Field after its collision with a stationary Hudson on 7 September.

An Airacobra strafes a No. 2 Ku Model 32 Zero trying to depart Buna on 26 August.

CHAPTER 12

TAINAN *KU* ANNIHILATED!
NEW GUINEA 23 AUGUST – 8 SEPTEMBER

Allied reconnaissance kept a keen eye on Lae and Buna, with 40 fighters noted at the former on 22 August and 30 fighters reported at Buna two days later. While these numbers were over-estimates and those seen at Lae included several non-operational airframes that had been there for some time, the Allies had nonetheless successfully detected the first deployment of Zeros to the newly operational airfield at Buna.

The Allied response was swift. An immediate strafing attack was ordered by Airacobras, as Buna was easily within the operational reach of P-39s from Port Moresby. However, the first attempts were frustrated by the ongoing bad weather which continued to scour the entire eastern half of New Guinea. It was not until first light on 26 August that eight P-400s from the 80th FS succeeded in taking off from Twelve-Mile, albeit in the face of heavy rain. Unfortunately, two fighters quickly aborted due to technical problems, including the squadron CO Captain Phil Greasely.

The six remaining Airacobras arrived over Buna at 0725 and commenced a steep diving approach. The attack could not have been timed better as they found two trios of Zeros making their take-off runs. These represented one *shotai* each from the Tainan and the No. 2 *Kokutai* who were operating closely together at Buna. Lieutenant Bill Brown and his wingman Lieutenant Danny Roberts each found easy targets in the first trio. Barely airborne, FPO1c Iwase Ki'ichi and FPO3c Ihara Daizo from the No. 2 *Ku* were shot down in flames and killed. The trailing Airacobras accounted for a third Zero from the other trio, flown by FPO3c Nakano Kiyoshi of the Tainan *Ku*.

Using their speed the Airacobras pulled up steeply and doubled back to engage in low-level combat. Of the remaining three Zeros two more sustained hits but managed to land. Warrant Officer Tsunoda Kazuo was uninjured, however his Model 32 tail code Q-102 was abandoned at Buna due to damage incurred from Airacobra guns. It was an unusual patriotic donation airframe too, with the fuselage marking HK-872 indicating his Zero had been donated by a Korean businessman. During this final pass Lieutenant Gerald Rogers' P-400, with British serial BW112, was hit by AA fire. After successfully ditching his fighter near a coastal village, Rogers was fortunate to avoid Japanese troops in the area. With the help of friendly natives he eventually returned to Port Moresby after a month-long ordeal in the jungle.

During the combat FPO2c Yamazaki Ichirobei was seriously wounded by gunfire. He managed to return to Buna where he was evacuated by ship to Rabaul. From there he would soon be repatriated back to Japan by Mavis flying boat for long-term medical treatment. Luck had finally run out for Yamazaki, for only three days previously he had force-landed at Buna, receiving

Model 32 Zero Q-102, flown by Warrant Officer Tsunoda Kazuo, which was abandoned at Buna after being shot up by Airacobras on 26 August.

only mild bruises. Until this fateful event, Yamazaki had enjoyed a reputation among Rabaul's fighter corps for being one of the blessed pilots in the Pacific – back in the early days of the No. 4 *Ku* he had survived a force-landing behind Lae when hit by Hudson gunfire, returning to base on a log raft. The effect on morale of his comrades is not recorded but his departure could hardly have lifted their spirits.

The Airacobras were followed by a formation of B-26s which bombed the southern part of Buna's runway and the dispersal area. Light damage from AA fire was sustained but all of the bombers returned safely. These successful raids on Buna meant that no Japanese fighters were present over Milne Bay on this important day, where the Japanese landing had just taken place. Taking advantage of uncontested skies, the Kittyhawks were able to carry out their devastating ground attack sorties.

The effort against Buna was repeated on 27 August by seven 19th BS Marauders and fourteen 41st FS Airacobras. The first combat air patrol over Buna, a *shotai* of three Tainan *Ku* Zeros led by Lieutenant Ono Takeyoshi, had been in the air for one and a half hours when the American intruders were sighted at 0720. This trio was first engaged by the Airacobras while the Marauders thought they had been attacked by a second patrol. The action was later described by an RAAF co-pilot on one of the B-26s, Flying Officer Robertson:

> Just as we arrived six Zeros started in to attack us but the 39s attacked on our right side and almost immediately one Zero crashed into the sea. So we bombed with no opposition. On turning out to sea after dropping we could see the 39s strafing the place and two ships. A Zero had a 39 on its tail and suddenly the Zero exploded and started to burn and went dull red and went straight in and ploughed under the water like a torpedo, kicking up great

FPO2c Yamazaki Ichirobei in the cockpit of his Model 21 at Lae. He survived three forced-landings in New Guinea before being repatriated to Japan for medical treatment after being wounded on 26 August.

spouts of water … It was the first pursuit [fighters] we have had on any raid and it certainly was the goods.

Despite Robertson's observations, and official awards to the Airacobra pilots for five Zeros downed, there was just one Japanese loss, Tainan *Ku* pilot FPO3c Matsuda Takeo. This combat exemplifies the odds which the Japanese faced during this collapsing campaign: a trio of three Zeros had taken on 21 Allied bombers and fighters, from which they claimed three Airacobras but in fact downed none. Combat barely lasted ten minutes before Ono and wingman FPO3c Uehara Sadao had returned to Buna, minus Matsuda. Each had fired 700 rounds of combined ordnance.

However, 27 August topped off a disastrous couple of days for the Buna fighter detachment as on this day four Zeros were lost over Milne Bay. Overall, eight pilots and their Zeros had been lost in 48 hours, plus unit favourite Yamazaki evacuated to Japan wounded and other fighters destroyed or damaged on the ground at Buna.

By this time, the Allies had excellent intelligence on the Buna airfield with an air photo interpretation report giving considerable detail on the installation. The main runway was 1,200 yards long with nineteen adjoining dispersal bays. Thirteen fighters were counted in the photo, four of which were damaged suggesting the photo had been taken after the first raid on 26 August. Several AA positions were noted comprising medium and heavy calibre AA guns as well as a series of machine gun positions. A second runway was east of the first one but unfinished and only 1,000 yards long.

Buna continued to be targeted as weather allowed, and B-26s escorted by 41st FS Airacobras attacked the airfield again on 29 August. Here they found some unusual victims in a pair of Mitsubishi G6M1-Ls which were transport versions of Betty bombers, operated by the Tainan *Kokutai* transport detachment. Both V-902 and V-903 were destroyed, leaving just a solitary example (V-901) in service with the Tainan *Ku*. The Airacobra pilots also reported leaving two Zeros burning, although these might have been non-operational airframes.

On the last day of the month, 31 August, a new type of Allied aircraft made its combat debut in

The wreckage of Tainan Ku transport Betty V-902 after it was destroyed at Buna by strafing Airacobras on 29 August.

the South Pacific, however behind this landmark event lies both a morale crisis and the seeds of low-level attack tactics which would revolutionise aerial warfare in the theatre. The aircraft were Douglas A-20As of the 8th BS whose crews had been converting to the type at Charters Towers, trying to take their minds off the disastrous A-24 losses of 29 July which had cost, inter alia, the life of their squadron CO Major Floyd "Buck" Rogers.

Engineering units at Townsville had been busy converting the A-20As into low-level strafers, and this mission was the first large-scale trial of the concept. A plan was developed whereby nine B-26s led by Lieutenant Walter Krell, each loaded with 30 x 100-pound bombs, would strike Lae airfield first from 5,500 feet at around midday. It was hoped that this first attack would render the defenders on the ground too dazed to fire back at thirteen low-level A-20A strafers, following one minute behind at treetop level. The plan worked with no combat losses, and the seeds had been sewn for the full-scale adoption of the concept of low-level attack aviation in the theatre.

For the return journey 80th FS Airacobras escorted the A-20As home, while the faster B-26s forged ahead by themselves. Over the Kokoda Pass the A-20A crews were shocked to see the pilot of a P-400 (BW147; squadron code "L") suddenly bail out. This was Lieutenant Charles Cobb who parachuted into mountainous jungle behind enemy lines after experiencing an unexpected engine failure. Initially given up as lost, Cobb reappeared in Port Moresby a month later with friendly locals.

Meanwhile the 8th BS A-20As, now forward-based at Port Moresby's Kila 'drome (Three-Mile), were reinforced on 2 and 8 September with two flights of six A-20As each. The second flight was from the 89th BS which had also been converting to the type. Unbeknownst to the men of the 8th BS, higher command was concerned with the low morale of the squadron ever since the loss of their core cadre on 29 July, and decided to re-assign the squadron full-time to maintenance and logistics duties to service the 3rd BG's other three squadrons. At Kila 'drome the squadron's aircrews were shocked when ordered to hand over all of their A-20As to the 89th BS.

Thus the 8th BS was effectively removed from combat until mid-1943, although many of their combat-tested crews were seconded during this interval to other squadrons. Eventually it would re-equip with B-25Ds and then A-20Gs before returning to full operational duty. Whilst the 8th BS contemplated their misfortune, in the first week of September the 90th BS received a dozen new B-25Ds with crews. Such reinforcements were indicative of the increasing industrial firepower becoming available to USAAF efforts in New Guinea. However, the attached crews were all inexperienced, having just completed flying training. It would take them several months of additional training and combat before they were to become effective. The squadron was assigned around twenty RAAF co-pilots to assist them in getting used to the theatre.

Meanwhile during the last week of August, the heavy bombers of the 19th BG continued operations against Rabaul. The first of these was a night raid, with eight Fortresses departing Seven-Mile at around 2200 on 24 August. This mission was affected by the same violent storms that bedevilled so many other operations at this time. After flying through the storms to bomb Vunakanau, the B-17s battled them on the way back with aircraft experiencing ice formation. Then they found the Port Moresby area closed in by low cloud down to only 400 feet, but all of the bombers had managed to land safely by 0500. Major Rouse wrote in his diary:

> Bad time finding it [Port Moresby]. Navigator really on the ball.

Seven Fortresses returned to Rabaul on 29 August for a daylight attack. Unusually, these were observed over Rabaul by a lone 435th BS B-17F on a photo reconnaissance mission which joined the formation as a defensive tactic. The B-17F was flown by Captain Fred Eaton and RAAF co-pilot Sergeant Merv Bell, an experienced duo who had flown many LB-30 reconnaissance sorties in recent weeks. Bell later wrote a detailed description of the 29 August mission:

> … as we neared Rabaul in clear sky we could see the harbour crammed with navy and merchant ships and could also see fighters taking off from the airfield at Vunakanau. We were at 28,000 feet with ack ack bursting around and below us. Just as we finished our final photo run over the harbour and three airstrips we sighted a group of seven Fortress aircraft coming in on a bombing run from the north on Vunakanau airstrip where there were many aircraft parked. They were some miles to our right. We decided to join them for greater firepower to fight off the Zeros that were now at our height. Two of them attempted to cut us from the strike force. The first did a stall turn from two o'clock and attacked with machine guns and cannon. We turned inside his turn. He fired one of the longest bursts I have ever experienced but did not make a hit. He flashed past our nose. Our turret and nose guns picked him off. We could see our tracers hit the Zero. He dived away burning and smoking.

His mate was now above our altitude at 12 o'clock. He swung to the right, dived below us and closed to fire at us from below. He had commenced to fire when the belly turret picked him off. The Zero blew to pieces. The other Zeros had headed for the strike force who were in close formation now, below us to the right. We gave our engines maximum power in a shallow dive. It took ten minutes to catch up to the strike force aircraft. By this time it was getting close to the Zero's endurance at max revs. The five of them broke off and headed for home. We stayed with the formation for another ten minutes, then broke away to complete our detail …

Although no Japanese fighters were lost, Bell's graphic recollections align closely to Tainan *Ku* records. In fact, only three Zeros were airborne. At 0910 Flyer2c Sasamoto Takamichi had departed Rabaul on patrol and about half an hour later reported eight incoming Fortresses. With many flyers away in Lae and Buna, Rabaul was bereft of experienced pilots, however *hikotaicho* Lieutenant-Commander Nakajima Tadashi had purposely left Warrant Officer Takatsuka Tora'ichi behind at Rabaul for such occasions. Takatsuka was the most experienced Warrant Officer pilot in the entire Tainan *Ku*, and also the most aggressive. It was he and Flyer2c Nagao Nobuto who launched after Sasamoto's radio call on seeing the Fortresses. Nagao and Sasamoto were both rookie pilots and had barely been in combat for two months. Movements of the combat are more complicated by the involvement of a fourth Japanese aircraft, a C5M Babs. The Babs was commanded by FPO2c Iwayama Takashi who dropped aerial bombs into the Fortress formation at 1015 and claimed one of the bombers as a "probable". This interception did not prevent the seven B-17s from unloading 40 bombs over Vunakanau at 1015.

The loss of Airacobra pilot Cobb as described above was followed by two more 80th FS losses within just a week. On 1 September a P-400 was lost after experiencing engine failure. The 41st FS pilot, Lieutenant George Helveston, bailed out of BX146 "Q" shortly after taking off from Durand 'drome. Five days later, on 6 September, the third 80th FS P-400 was lost when Captain Francis Potts bailed out after being hit by ground fire while dive-bombing Japanese positions near Myola. Potts was possibly killed after bailing out at very low altitude, however there is another more likely fate. Several Australian soldiers saw him land safely on the side of a distant creek bed. However, he had landed behind Japanese lines, and if so, it is also possible he was captured and killed. At any rate, he remains missing in action.

On 2 September a formation of B-26s bombed Lae from 9,300-feet, dropping 166 x 100-pound bombs about half of which landed across the runway. On the return flight the bombers were met by a welcome but unnecessary escort of Airacobras.

Meanwhile, high in the Owen Stanley mountains the Japanese continued to push Australian forces back along the Kokoka Track. A heavy engagement fought at Isurava between 26 and 31 August became the first major battle of the campaign. Although reinforced by newly arrived AIF battalions, the result was the Australians continuing to retreat. In fact, following the failure to capture Milne Bay and the rushing of available IJA troops to Guadalcanal, the Japanese had deferred their intention to advance on Port Moresby overland. Instead those forces in Papua were ordered to capture a holding position in the mountains from where a future offensive might be launched with fresh troops.

However, these intentions were not known to the Allied commanders in New Guinea, for whom the evolving Kokoda Track campaign seemed to bring daily new emergencies. Following the major battle at Isurava, the Australians fought a number of delaying actions but retreated to

Douglas A-20A, 40-109, Spook 89th BS, 3rd BG, Kila 'drome, early September 1942.

the southern slopes of the Owen Stanleys and were forced to abandon their aerial supply base at Myola.

At this time several frantic air support missions were flown, including the one which cost the life of Potts on 6 September as noted above. Given the thick jungle canopy and lack of readily available landmarks, the missions were usually ineffective. An exception was on 7 September when eight B-26s bombed and strafed Japanese positions accurately, causing several confirmed casualties. The history of the 3/144th Infantry Battalion states:

> Lieutenant Kazue, the unpopular 3/144 pay officer, and ten others were killed.

Meanwhile small numbers of Tainan and No. 2 *Kokutai* Zeros were flying convoy protection patrols from Buna in early September, typically involving just two or four aircraft. At this time another convoy brought fresh troops to Buna, with the ships undetected by Allied reconnaissance that was fully occupied at Milne Bay.

Over Rabaul on 1 September, the Tainan *Ku* managed to scramble just a single Zero at 1020 to try and intercept a 63rd Bombardment Squadron B-17F on a reconnaissance mission (the 63rd was part of the 43rd BG and had recently commenced operations from Mareeba under the guidance of the experienced 19th BG). As described in the previous chapter, on 2 September the Tainan *Ku* could only muster only a pair of Zeros (alongside several from the No. 2 *Ku*) for a strike launched against Milne Bay.

The fortunes of the once mighty Tainan *Ku* had indeed fallen to new lows, particularly in light of the loss of some of their most experienced pilots over Guadalcanal. In late August a Japanese diarist at Buna wrote:

> Tainan *Kokutai* has been annihilated. The commander has gone back to Rabaul.

Such strong language was certainly true of the forward deployments sent to Lae and Buna. However, there was still a potent rump of pilots from which to draw at Rabaul, and on 2 September ten Chitose *Ku* Zeros were flown to Rabaul where they were reassigned to the Tainan *Ku*, effectively reinvigorating the unit.

A curious native boy is shown the front end of a Tainan Ku Zero while a No. 4 Ku Betty is on finals to Lakunai. Fighter strength at Rabaul was reduced to critical levels in early September with only one Zero available to intercept a B-17 at the start of the month.

The reconnaissance arm of the Tainan *Ku* was struggling too, with only occasional missions managed by the handful of operational C5M Babs. One of these flew over Buna via Gasmata on both 2 and 6 September. Indeed, replacement of these useful aircraft was problematic as only a very small number had been built. IJN fleet records noted that operational attrition for the land-based reconnaissance planes was:

> ... very heavy and replacement was very difficult, thus causing considerable troubles in the operations.

Ultimately an effective reconnaissance solution would materialise in the shape of the sleek Ki-46 twin-engined Dinah, operated by the Japanese Army Air Force. The first of these would arrive in Rabaul in October, and the type was so successful it would also be adopted by the IJN.

The B-17 campaign against Rabaul resumed on 5 September with a strike by nine 19th BG Fortresses, however the day was strategically important when at Townsville Brigadier General Kenneth Walker was appointed Commander of Fifth Bomber Command. Three days later he issued orders that the 19th BG was to keep a full-time strike force of nine bombers at Port Moresby, with the remainder of the unit to be on standby at Mareeba.

On 6 September the IJN planned a sizeable raid on Port Moresby. An impressive 27 Bettys were assembled but to do this one *chutai* was from the Misawa *Ku*, another was from the Kisarazu *Ku* while the third was a combined *chutai* using aircraft from the Kisarazu and Chitose *Ku*. The escort was also combined, with a dozen Zeros from the Tainan *Ku*, nine from the No. 2 *Ku* and nine from the No. 6 *Ku*. This impressive formation did not get far in the face of bad weather.

Mitsubishi G4M1 Model 11 Betty, R-310, Kisarazu Ku, Vunakanau, September 1942.

After 40 minutes most of the aircraft had returned to Rabaul, with the exception of the No. 6 *Ku* Zeros which found their way to Lae instead.

The mission was rescheduled for the following day, 7 September, this time with the precaution of a C5M Babs flying ahead to check the weather. From 0800 Bettys from the three different units began taking to the air, eventually combining in a formation of 26 bombers from the Chitose, Misawa and Kisarazu *Kokutai*. These were escorted by an identical number of Zeros from the Tainan and No. 6 *Kokutai*. This would mark the first and last appearance of the No. 6 *Ku* over Port Moresby; the unit having only arrived at Rabaul on 21 August.

On this occasion the coordination between the fighters and bombers was less than ideal, with the bombers arriving over the target area before the fighters. However, this was of no consequence as those on the ground only received brief warning of the raid and there was no chance for Airacobras to make contact with the intruders. Bombs were dropped over Seven-Mile, where two P-400s were recorded as "badly damaged" and various others were "slightly damaged".

On the return flight one Kisarazu *Ku* Betty crash-landed on the northern coast of Papua, presumably as a result of damage from AA fire over the target zone. Some Bettys diverted to Lae on the way home due to excessive fuel consumption, but all of the other aircraft returned home safely. This mission was something of a landmark, being the final time that the Tainan *Ku* would appear over the skies of Port Moresby. During a six-month period beginning at the start of April 1942 the Tainan *Ku* had routinely appeared over the Papuan capital and had been the nemesis of Allied fighter pilots in the months up to July. However, attrition had taken a toll on the best Tainan *Ku* pilots which could not easily be replaced, and by late September the crack unit had become a shadow of its former self, reflecting Japanese fortunes generally.

Also on 7 September, the few remaining No. 2 *Ku* Zeros at Buna were withdrawn permanently back to Lae, effectively ending the brief use of the newly opened airfield.

Late on the morning of 29 August the 1,883-ton Burns Philp motor vessel *Malaita* steamed out of Port Moresby's harbour after having arrived from Townsville six days earlier with a cargo of troops and supplies. *Malaita* was escorted by the destroyer *Arunta*. Just after exiting the passage through the outer reef, *Malaita* was torpedoed by the submarine *RO-33* which as noted in Chapter 8 had sunk the *Mumutu* in the Gulf of Papua some three weeks earlier. Subsequently the *Arunta* acquired a sonar contact and carried out depth-charge attacks for 90 minutes. Oil and bubbles rising to the surface gave rise to the belief that a submarine had been sunk, and this was indeed so: *RO-33* was lost with all hands. The *Malaita* remained afloat and was towed into Port Moresby harbour. After some temporary repairs the motor vessel was able to sail for Australia on 14 September.

Finally, a recap of RAAF Catalina operations during the period 23 August to 8 September is in order. The submarine activity in the Gulf of Papua and the importance of supplies getting to Milne Bay resulted in many convoy protection missions being flown. In addition, some of the nuisance night raids were continued. Four Catalinas (A24-16, -24, -25 and -27) bombed Buka's airfield on the night of 28-29 August. This effort was repeated by three Catalinas (A24-24, -26 and -27) on the night of 1-2 September, and again by two Catalinas (A24-10 and -16) on the night of 2-3 September. During the final raid the crew of A24-16 reported hitting a fuel dump and starting a large fire. A variation to these raids was when A24-1 dropped supplies to coastwatchers on New Ireland on the night of 26-27 August.

The Australian destroyer HMAS Arunta seen on trials off the New South Wales coast in late April 1942. A few months later the Arunta was actively involved in escorting supply ships into Milne Bay and sunk the submarine RO-33 off Port Moresby on 29 August.

CHAPTER 13

CONCLUSION

Each of the first three volumes of this series chronicled periods in which the intervention by carrier-based airpower, from both the Allied and Japanese sides, had a decisive effect in the New Guinea area. The period chronicled by this fourth volume, 19 June – 8 September 1942, was the first such period where carriers were absent from New Guinea. This resulted in new and different key trends developing both in the land-based air war and also in terms of general strategy.

Despite their loss at the Battle of Midway in June, the Japanese entered this period full of confidence that their strategy of territorial expansionism was still both valid and effective. Indeed, for the weeks up until the end of July they had good reason to be optimistic. Their airfield construction efforts on Guadalcanal were continuing largely unhindered and they had successfully established a firm beachhead in the Buna-Gona area. Subsequent convoys landed a strong force of IJA troops in northern Papua such that the new campaign underway over the Owen Stanley mountains held much promise. In addition, airfield construction was underway at Buna and plans were underway to land naval forces at Samarai to assist with the advance on Port Moresby.

However, in early August the Japanese suffered two great and unexpected shocks which forced them from taking the initiative to being reactive to Allied moves. The first of these was the discovery of the Allied base at Milne Bay on 3 August, where a large construction effort had been undertaken in secret. This was just a short distance from the planned Japanese landing at Samarai. The second shock was the massive American invasion of Tulagi and Guadalcanal just four days later, just when the IJN was on the cusp of moving land-based air forces to the newly built airfield at Lunga Point. While the Japanese had some inkling of an imminent Allied counter-offensive, the scale and audacity of the landing was unforeseen, sudden and devastating.

The two great shocks at Milne Bay and Guadalcanal represented a perfect storm for the Japanese, who in view of their established landing at Buna, now faced great demands on three fronts. The assortment of persistent and unwelcome news exercised robust discussion among Rabaul's commanders which at times unhappily verged on the confrontational. Indeed, the new revelations caused considerable flurry and late nights in Rabaul's various headquarters that arguably resulted in some poorly planned knee-jerk decisions. However, options were limited, especially from an airpower perspective, and commanders lacked detailed intelligence about the Allied forces they were facing.

Soon a vast quantity of resources would be expended on trying to recover Guadalcanal, which became the number one priority for Mikawa in Rabaul. However, the attempt by the IJN to capture Milne Bay still proceeded in late August, despite the fact the Japanese lacked vital intelligence about their objective and that the weather at that time of year was seasonally bad. Then there was the ongoing need to run convoys to Buna and maintain the troops fighting on the Kokoda front. All of these commitments spiralled rapidly and forced the need for IJN / IJA collaboration at every turn. This was a hectic period in Rabaul with plans being cast then restructured almost daily.

When the Japanese evacuated Milne Bay at the end of the first week of September, the outcomes of the campaigns in Papua and Guadalcanal were still far from certain. In Allied headquarters, both in the SWPA and the SoPA, this was also a time of great emergency. The demand for airpower was unprecedented in the theatre, not only as a weapon against enemy air forces but also in direct support of ground troops.

Yet for much of this period the burden fell to the same tired units continuing to wage war. From the Japanese side the herculean efforts of the Tainan and the No. 4 *Kokutai* continued, and two achievements stand out. The first was the ongoing air campaign against Port Moresby which suffered 76 raids up until the start of August. Although significant damage was inflicted on only a few occasions, the sustained nature of this campaign meant that no Allied medium or heavy bombers were permanently based in Port Moresby. This effectively blunted the effectiveness of the Allied bomber force by a factor of at least 50%, probably even greater.

The second factor was the tactics of Tainan *Ku* Zeros as a defensive force. The far-ranging American B-17s were both feared and respected in their ability to strike at will against almost any target in the region, weather permitting. Since the start of the Pacific War until mid-June, only six Fortresses had been lost in the South Pacific and in only one of these losses was aerial combat with Zeros a possible contributing factor. However, during August Zeros accounted for five definite B-17 losses, in a small but significant victory for the Japanese fighter force. Following hard-won experience, the Tainan *Ku* Zero pilots concentrated on frontal attacks in pairs or even by single aircraft. This tactic proved effective and minimised their own losses as the Japanese faced only the Fortress' relatively weak front facing armament. They were also better prepared to deal with the B-17's large size as a target, which in previous months had repeatedly led pilots to open fire from too great a range.

However, there were many failures of Japanese airpower during this period too. After sinking the *Macdhui* on 18 June, Japanese aircraft failed to land any further strikes on the soft underbelly of merchant ships supplying both Port Moresby and Milne Bay. Also, it was a twin failure of aerial reconnaissance to either detect early construction activity at Milne Bay or to spot the massive incoming Guadalcanal invasion convoy, although weather was a factor in both cases. Lastly, as Allied airpower was beginning to view low-level attack as a promising future strategy, the concept was not properly understood by the Japanese. During their low-level attack against Milne Bay on 27 August the lightly built IJN aircraft without fuel tank protection fared poorly, with six Zeros and Vals lost. Likewise, the low-level efforts of Bettys and Vals against the invasion fleet off Guadalcanal on 7 August were catastrophic.

Further, by September the veteran units of the Tainan and the No. 4 *Kokutai* were seriously depleted both numerically and qualitatively. Guadalcanal in particular ensured a terrible acceleration in the loss rates of their most experienced aircrews. Fortunately for the Japanese a broad collection of replacement units had been earmarked for the theatre, including the No. 2, No. 6, Chitose, Misawa and Kisarazu *Kokutai*. However, some of the incumbent aircraft types were less than ideal for South Pacific combat. The Val dive-bombers lacked range and could only carry light bombs, while the new Model 32 Zeros lacked the endurance for long range escort and more distant offensive missions.

Perhaps surprisingly, Japanese losses during this period were relatively light: just 37 aircraft and 46 crewmen (see the Appendices for a detailed listing of losses from both sides). This compares with comparatively heavy losses from the Allied side: 89 aircraft and a staggering 191 crewmen. Noteworthy losses were some 19 B-17s, compared to only six lost in the previous periods. However, the majority of these losses were due to accidents, many of them weather related.

Arguably the most effective contribution made by airpower during this period was the low-level fighter attacks made by both the USAAF Airacobras and the RAAF Kittyhawks. These attacks greatly assisted the successful defence of Milne Bay and also led to the evacuation of Buna by the force of Zeros and Vals briefly based there. This was an innovative yet practical use of these two types, which were relatively heavily armed and gave their best performance at low to medium altitude. It contrasted with the forlorn attempts by these fighters to intercept bombers high over Port Moresby only to face Zeros with a high-altitude performance advantage.

In contrast the efforts of the bomber force, mainly that by B-25s, B-26s and B-17s, paid few real dividends despite the extraordinary claims of General Kenney. The fact was bombing from medium and high altitudes had proved largely ineffectual in New Guinea conditions. On occasion, such as when B-17s attacked from low altitudes, some results were observed such as the sinking of the *Ayatosan Maru* off Gona, but such occasions were still the exception rather than the rule.

While neither side possessed significant power to deliver knock-out blows, two Allied initiatives towards the end of this period envisaged the potential of future hitting power. The first of these developments was the combat debut of 3rd BG A-20A strafers on 31 August, and the second was the decision to begin permanently basing detachments of bombers, including A-20s and B-17s, at Port Moresby. This call forward was a major win for the Allies and followed a slackening of Japanese air raids against the location, given the needs of other fronts over Papua and Guadalcanal.

In line with these developments was a change in command arrangements which would ultimately permit a steady increase in Allied strength and an increasingly aggressive approach. In early September all USAAF units in the SWPA were reorganised under General Kenney's newly created Fifth Air Force. The RAAF too reorganised its modest but significant New Guinea forces under No. 9 Operational Group, and Milne Bay rather than Port Moresby became the centre for RAAF operations in the theatre. Importantly, the RAAF now fell under the umbrella of Fifth Air Force command, thus ensuring a coordinated response.

Meanwhile the biggest anxiety for Allied command was the burgeoning requirement to support the ground forces, both in terms of transport and direct air support. The number of transport aircraft in Port Moresby was pitifully deficient to support the Australian land forces fighting in the Owen Stanleys. This was particularly true after the 17 August Japanese raid on Seven-Mile which removed several invaluable transports from service. It was also true that a large number of aircraft was needed to intensify efforts during the short periods that clear weather permitted operations.

Some much-heralded efforts were also made in direct support of the land forces, especially by B-26s and P-39s. However, it was not known at the time how ineffective these efforts were. Such missions needed to be flown with the aid of dedicated and direct radio communication with ground forces, but this doctrine lay in the future.

The end of this period of fighting saw the Allies best positioned to seize the initiative in the New Guinea air war. However, the Allied forces had suffered by far the heaviest losses in the June-September period, and the Japanese were in the process of sending several new air units to Rabaul. Also, the Japanese still had uncontested naval supremacy in New Guinea waters together with the promise of strong IJA reinforcements. The end of this volume sees the New Guinea air war reach an exponential level of complexity compared to just three months earlier. Yet, no side had thus far established a clear advantage over the other, and Japanese fortunes in New Guinea would be increasingly drained and distracted by the evolving quagmire in the neighbouring Solomons.

APPENDICES

APPENDIX 1 – ALLIED AIRCRAFT LOSSES & FATALITIES
Confirmed Allied Military Aircraft Losses
19 June to 8 September 1942*

From 7 August losses in the Solomons are excluded.

* four North Queensland losses from 4 – 12 June that were excluded from Volume 3 have been included here. Note as a general rule, accidents in North Queensland not directly involved with operational flying have been excluded from these tables. However, such distinctions are often not clear-cut and have been left to the discretion of the authors.

	DATE	TYPE	SERIAL	UNIT	COMMENTS	FATALITIES*
1	4 Jun 42	CAC Wirraway	A20-494	No. 24 Sqn	Wirraway A20-494 of No. 24 Squadron, RAAF, struck a dispersal bay during a night take-off from Mackay and was wrecked.	
2	5 Jun 42	Beaufort	A9-50	No. 100 Sqn	Accident at Mareeba; written-off.	
3	5 Jun 42	Beaufort	A9-40	No. 100 Sqn	Accident at Coen; written-off.	
4	12 Jun 42	Beaufort	A9-56	No. 100 Sqn	At 0830 on 12 June 1942 Beaufort A9-56 of No. 100 Squadron, RAAF, departed Mareeba to fly an anti-submarine patrol off the North Queensland coast. The aircraft never returned and remains missing with the loss of all four crew.	4 killed
5	23 Jun 42	DC-2		21st TCS	Collided with a truck during landing at Charters Towers and written off. Originally NEIAF registration PK-AFK.	
6	25 Jun 42	P-39F Airacobra	41-7271	39th FS	Written off during crash landing at Twelve-Mile after combat with Zeros. Lt Robert Rose OK.	
7	26 Jun 42	Beaufort	A9-52	No. 100 Sqn	After night raid against Salamaua got lost in bad weather on return flight to Port Moresby. S/L Charles Sage and three crewmen killed. Wreck found in 1987 at 7,000-feet in mountains north-west of Port Moresby.	4 killed
8	26 Jun 42	P-39F Airacobra	41-7137	40th FS	P-39 flown by Lt William Stauter missing after combat with Zeros over Port Moresby.	1 killed
9	26 Jun 42	CAC Wirraway	A20-84	No. 24 Sqn	Wirraway A20-84 of No. 24 Squadron crashed into Mount Louisa, not far from RAAF Townsville, with the loss of both crewmen.	2 killed
10	4 Jul 42	Hudson	A16-193	No. 32 Sqn	Departed Horn Island on the evening of 3 July for a night raid on Salamaua but never returned. Pilot Flight Lt Pat McDonnell and three crew missing.	4 killed
11	4 Jul 42	P-39F Airacobra	41-7148	39th FS	Lt James Foster bailed out safely after combat with Zeros over Port Moresby.	
12	4 Jul 42	P-400 Airacobra	AP 378	39th FS	Lt Frank Angier bailed out safely after combat with Zeros over Port Moresby.	
13	4 Jul 42	P-400 Airacobra	BX 180	39th FS	Lt William Marlott made a forced landing near the coastal village of Boera after combat with Zeros over Port Moresby.	
14	4 Jul 42	B-26 Marauder	40-1468	33rd BS	Lost during bombing mission against Lae after mid-air collision with Zero. Pilot Lt Milton Johnson and six crewmen killed.	7 killed

	DATE	TYPE	SERIAL	UNIT	COMMENTS	FATALITIES*
15	5 Jul 42	B-26 Marauder	40-1497	408th BS	Destroyed on ground at Port Moresby by Japanese air raid.	
16	5 Jul 42	B-26 Marauder	40-1510	2nd BS	Destroyed on ground at Port Moresby by Japanese air raid.	
17	5 Jul 42	P-39 Airacobra	BW 161	40th FS	Destroyed on ground at Port Moresby by Japanese air raid.	
18	5 Jul 42	B-26 Marauder	40-1533	33rd BS	Written off in a landing accident at Cairns following an engine failure. Had just returned from New Guinea and was en route from Cooktown.	
19	6 Jul 42	P-400 Airacobra	AP 377	40th FS	Lt Howard Welker killed after bailing out at low altitude near Gona Mission, following combat with Zeros.	1 killed
20	11 Jul 42	P-400 Airacobra		40th FS	Broke up during high speed forced landing near 30-Mile following engine failure at 15,000-feet. Lt Edward Gignac seriously injured.	
21	11 Jul 42	P-400 Airacobra	BW 117	40th FS	Lt Orville Kirtland missing after combat with Zeros over Port Moresby.	1 killed
22	14 Jul 42	B-17 Flying Fortress	41-2655	30th BS	Forced down in ocean soon after take-off from Horn Island in stormy conditions. Three crew killed.	3 killed
23	14 Jul 42	B-17 Flying Fortress	41-2636	30th BS	Forced down in ocean soon after take-off from Horn Island in stormy conditions.	
24	14 Jul 42	Lockheed Lodestar	VHCAD (ex LT-914)	22nd TCS	Missing during flight from Townsville to Cooktown amidst low cloud. Pilot Lieutenant Robert Davis and three passengers missing. Formerly NEIAF LT-914.	4 killed
25	16 Jul 42	B-17 Flying Fortress GI Issue	41-2421	435th BS	Crashed during night landing accident at Horn Island. Pilot Major Clarence "Sandy" McPherson and sixteen others onboard killed.	17 killed
26	18 Jul 42	Hudson	A16-179	No. 32 Sqn	Destroyed on ground at Port Moresby by Japanese air raid.	
27	18 Jul 42	P-39D Airacobra	41-6783	39th FS	Lt Frank Angier bailed out after becoming lost in bad weather on return from reconnaissance mission to Lae. Angier eventually returned to Port Moresby.	
28	22 Jul 42	P-39D Airacobra	41-7143	80th FS	Lt David "Pinky" Hunter ditched near Gona after being hit by AA fire. Captured and taken to Rabaul where he was later executed.	1 killed
29	22 Jul 42	P-39D Airacobra Papuan Panic	41-38353	40th FS	Lt Garth Cottam missing during attack on Buna in bad weather.	
30	22 Jul 42	Hudson	A16-201	No. 32 Sqn	Crashed in north Papua after being intercepted over the sea by Zeros and shot down. Pilot Officer Warren Cowan and three crew killed.	4 killed
31	25 Jul 42	P-400 Airacobra		40th FS	Lt Frank Beeson shot down by Zeros near Buna during fighter-bomber mission.	1 killed
32	25 Jul 42	P-400 Airacobra		40th FS	Lt David Hoyer wounded after combat with Zeros. Managed to land his damaged Airacobra at Port Moresby where it was written off.	
33	26 Jul 42	B-25C Mitchell Aurora	41-12792	13th BS	Shot down by Zeros near Buna. Capt Frank Bender and one other crewman bailed out and survived.	4 killed
34	26 Jul 42	B-25C Mitchell	41-12470	90th BS	Shot down by Zeros near Buna flown by Lt Ralph Schmidt.	5 killed
35	27 Jul 42	B-17E Flying Fortress	41-2460 Flying Dutchman	30th BS	Wrecked after collision with 41-2640 Tojo's Physic on the ground at Horn Island.	

	DATE	TYPE	SERIAL	UNIT	COMMENTS	FATALITIES*
36	27 Jul 42	B-17E Flying Fortress	41-2640 *Tojo's Physic*	30th BS	Badly damaged after collision with 41-2460 on the ground at Horn Island. Subsequently flown to Mareeba where it was written off and used for spares.	
37	29 Jul 42	A-24 Banshee	41-15797	8th BS	Shot down by Zeros during raid on Buna and crashed into ocean. Major Floyd "Buck" Rogers and rear gunner Cpl Robert Nichols killed.	2 killed
38	29 Jul 42	A-24 Banshee	41-15819	8th BS	Hit by AA fire and crashed on beach near Buna with the loss of Captain Virgil Schwab and rear gunner Sgt Philip Childs.	2 killed
39	29 Jul 42	A-24 Banshee	41-15766	8th BS	Lt Claude Dean and gunner bailed out; captured and executed.	2 killed
40	29 Jul 42	A-24 Banshee	41-15751	8th BS	Lt Joseph Parker and gunner bailed out; captured an executed.	2 killed
41	29 Jul 42	A-24 Banshee	41-15798	8th BS	Lt Robert Cassels and gunner shot down; fate unclear possibly captured and executed.	2 killed
42	2 Aug 42	B-17E Flying Fortress	41-2435	28th BS	Intercepted by Zeros and shot down near Cape Ward Hunt. Only 1 crewman bailed out and survived. Lt William Watson and rest of crew killed.	8 killed
43	2 Aug 42	P-400 Airacobra	AP 290	41st FS	Lt Jesse Dore shot down by Zeros near Buna.	1 killed
44	2 Aug 42	P-400 Airacobra	AP 232	41st FS	Lt Jesse Hague survived after being shot down by Zeros near Buna. Seen on ground exchanging fire with Japanese forces but fate remains obscure and is missing.	1 killed
45	2 Aug 42	B-17E Flying Fortress *Uncle Biff*	41-9155	42nd BS	Written off after collision with another B-17 on the ground in the New Hebrides.	
46	4 Aug 42	B-17E Flying Fortress	41-9218	26th BS	Crashed during raid on Guadalcanal after mid-air collision with a Rufe. Lt Rush McDonald and crew killed.	9 killed
47	4 Aug 42	P-39F Airacobra	41-7165	39th FS	Major Jack Berry killed during practice dive-bombing of wreck off Port Moresby harbour.	1 killed
48	6 Aug 42	B-17E Flying Fortress	41-9221	42nd BS	Captain Rolle Stone ditched his Fortress off the northern end of Espirito Santo in darkness. Crew OK.	
49	7 Aug 42	B-17E Flying Fortress	41-2426	431st BS	Fortress flown by Major Marion Pharr lost in unknown circumstances over the Solomons. Possibly a friendly fire incident.	10 killed
50	7 Aug 42	B-17E Flying Fortress	41-9224 *Kai-O-Keleiwa*	98th BS	Lost in bad weather in vicinity of northern New Caledonia. Lt Robert Loder and crew lost.	9 killed
51	7 Aug 42	B-17E Flying Fortress	41-2617	30th BS	Crashed during take-off from Port Moresby and written off.	
52	7 Aug 42	B-17E Flying Fortress	41-2429 *Why Don't We Do This More Often?*	93rd BS	Shot down by Zeros during raid against Rabaul. Captain Harl Pease and one other crewman bailed out safely but were later executed by the Japanese.	9 killed
53	7 Aug 42	B-26 Marauder	40-1521 *Yankee Clipper*	19th BS	Crew bailed out after failing to find Port Moresby in bad weather. Lt Seffern and all onboard except two crewmen eventually returned safely to Port Moresby.	2 killed
54	7 Aug 42	B-26 Marauder	40-1496 *DIXIE*	19th BS	Force-landed in northern Papua after failed to find Port Moresby in bad weather. Lt Robert Hatch and crew were eventually returned to Port Moresby.	
55	7 Aug 42	P-40E Kittyhawk	A29-81	No. 76 Sqn	Written off in landing accident on Goodenough Island.	

	DATE	TYPE	SERIAL	UNIT	COMMENTS	FATALITIES*
56	8 Aug 42	Empire flying boat	A18-11	No. 33 Sqn	Destroyed after landing in heavy seas near survivors of the *Mamutu* in the Gulf of Papua. 1 crewman was killed, others eventually reached land safely in a rubber dinghy.	1 killed
57	9 Aug 42	B-17E Flying Fortress	41-2643	93rd BS	Shot down by Zeros during raid against Rabaul. Lieutenant Hugh Grundmann and crew killed.	9 killed
58	9 Aug 42	B-17E Flying Fortress	41-2452	93rd BS	Damaged during combat with Zeros over Rabaul. Ditched off Malapla Island near Milne Bay. Crew OK.	
59	11 Aug 42	P-40E Kittyhawk	A29-123	No. 75 Sqn	Shot down by Zeros, wreck found 25 miles east of Milne Bay base. Flying Officer Mark Sheldon killed.	1 killed
60	11 Aug 42	P-40E Kittyhawk	A29-100	No. 75 Sqn	Shot down by Zeros over Milne Bay. Warrant Officer Francis Shelley killed.	1 killed
61	11 Aug 42	P-40E Kittyhawk	A29-93	No. 75 Sqn	Shot down by Zeros over Milne Bay. Flying Officer Albert McLeod killed.	1 killed
62	11 Aug 42	P-40E Kittyhawk	A29-84	No. 76 Sqn	Damaged during combat with Zeros over Milne Bay. Flight Sergeant George Inkster bailed out at very low altitude over No. 1 Strip and was killed.	1 killed
63	13 Aug 42	B-26 Marauder	40-1492 *Sally Rand*	2nd BS	Lieutenant Harry Patteson ditched *Sally Rand* after the starboard engine was shot away by Zeros over Buna. The crew were rescued from Porlock Harbour by an RAAF Catalina. 1 crewman killed.	1 killed
64	13 Aug 42	F-4 Lightning	41-2125	8th PRS	Crashed on Misima Island in darkness after flying past Port Moresby from Horn Island. Pilot Lieutenant Paul Staller killed.	1 killed
65	14 Aug 42	B-17E Flying Fortress	41-2656 *Chief of Seattle*	435th BS	Shot down by Zeros in vicinity of Buna; flown by Lt Wilson Cook while en route for reconnaissance of Gasmata.	10 killed
66	16 Aug 42	P-40E Kittyhawk	A29-78	No. 76 Sqn	Written off after a tyre burst on take-off from Milne Bay and collided with a Hudson.	
67	16 Aug 42	Hudson	A16-218	No. 32 Sqn	Written off while waiting to take-off at Milne Bay and a Kittyhawk collided with it. One Hudson crewman killed.	1 killed
68	16 Aug 42	B-17E Flying Fortress	41-2434	30th BS	Crashed near Cairns after an accident with a flare caused an internal fire. Major Dean Hoevet and eleven passengers and crew killed.	12 killed
69	17 Aug 42	B-26 Marauder	40-1437 *Shamrock*	HQ Sqn	Destroyed on ground at Port Moresby during air raid.	
70	17 Aug 42	B-26 Marauder	40-1399 *The Avenger*	2nd BS	Written off after being damaged by shrapnel from Japanese bombs while taxying at Port Moresby. The RAAF co-pilot, Sergeant William Logan, was killed.	1 killed
71	17 Aug 42	DC-5	VHCXA		Destroyed on ground at Port Moresby during air raid.	
72	17 Aug 42	Lockheed Lodestar	VHCAG		Destroyed on ground at Port Moresby during air raid.	
73	17 Aug 42	Lockheed Lodestar	VHCAI		Destroyed on ground at Port Moresby during air raid.	
74	26 Aug 42	B-17F Flying Fortress	41-24354	93rd BS	Hit by AA fire from ships in Milne Bay and crashed into ocean. Captain Clyde Webb and crew killed.	9 killed
75	26 Aug 42	B-17E Flying Fortress	41-2621 *The Daylight Ltd*	93rd BS	Hit by AA fire from ships in Milne bay and written off after crash landing at Mareeba.	
76	26 Aug 42	P-40E Kittyhawk	A29-110	No. 75 Sqn	Ditched by Flying Officer Alan Whetters near a jetty on Sideia Island off Milne Bay's eastern-most cape.	
77	26 Aug 42	P-400 Airacobra	BW112	80th FS	Hit by AA fire over Buna and ditched. Pilot Lieutenant Gerald Rogers OK.	

	DATE	TYPE	SERIAL	UNIT	COMMENTS	FATALITIES*
78	27 Aug 42	LB-30 Liberator	AL515 *Yard Bird*	435th BS	Strafed and destroyed by Zeros at Milne Bay. Had already been damaged in a landing accident.	
79	27 Aug 42	P-40E Kittyhawk	A29-108 *Schuftie*	No. 75 Sqn	Crashed into jungle near Milne Bay, likely after combat with Zeros. Pilot Officer Stuart Munro killed.	1 killed
80	27 Aug 42	P-40E Kittyhawk	A29-92	No. 76 sqn	Crashed into jungle near Milne Bay, likely hit by small arms fire. Squadron Leader Peter Turnbull killed.	1 killed
81	28 Aug 42	P-40E Kittyhawk	A29-109	No. 75 Sqn	Crashed into hill near Seven-Mile while making landing approach in low cloud. Sergeant William Cowe killed.	1 killed
82	29 Aug 42	P-40E Kittyhawk	A29-106	No. 76 Sqn	Destroyed during night landing accident at Milne Bay. Ferry pilot Pilot Officer Brendon Davis killed.	1 killed
83	31 Aug 42	P-400 Airacobra	BX147	80th FS	Pilot bailed out near Kokoda while escorting A-20s. Lieutenant Charles Cobb OK.	
84	1 Sep 42	P-400 Airacobra	BX146	41st FS	Pilot bailed out following engine failure after take-off from Durand. Lt George Helveston OK.	
85	3 Sep 42	Hudson	A16-220	No. 6 Sqn	Crashed into ocean near Port Moresby. Pilot Officer William Campbell and three crew killed.	4 killed
86	4 Sep 42	B-25 Mitchell	*The Queen*	13th BS	Ditched in darkness while returning to Port Moresby from Milne Bay. Captain Gustave Heiss and crew were killed.	5 killed
87	4 Sep 42	B-25 Mitchell	*Hell Cat*	13th BS	Ditched in darkness while returning to Port Moresby from Milne Bay. Lieutenant Hubert Rapp and four crew killed.	5 killed
88	6 Sep 42	P-400 Airacobra	AP359	80th FS	Hit by AA fire near Myola. Captain Francis Potts killed.	1 killed
89	7 Sep 42	Beaufighter	A19-13	No. 30 Sqn	Slid off runway during take off from Milne Bay and written off after collision with Hudson.	
					* Total Fatalities (note other fatalities occurred in aircraft not classed as lost)	191

Breakdown of aircraft lost:

P-39 / P-400 Airacobra	21
B-17 Flying Fortress	19
B-25 Mitchell	4
B-26 Marauder	9
A-24 Banshee	5
Lodestar	3
Hudson	5
P-40E Kittyhawk	11
Bristol Beaufort	4
CAC Wirraway	2
Other	6

(DC-2, DC-5, Empire flying boat,
F-4 Lightning, Beaufighter,
LB-30 Liberator)

89

APPENDIX 2 – JAPANESE AIRCRAFT LOSSES & FATALITIES
Confirmed Japanese Military Aircraft Losses
19 June to 8 September 1942*

from 7 August onwards losses in the Solomons are excluded.

	DATE	TYPE	SERIAL/ UNIT	COMMENTS	FATALITIES*
1	26 Jun 42	Nell	Genzan *Ku*	Badly damaged in combat with Airacobras after bombing Port Moresby. Eight crew were wounded, pilot FPO1c Hirayama Jinro made a forced landing at Lae. Aircraft written off.	
2	4 Jul 42	Zero	Tainan *Ku*	Lost while defending Lae after mid-air collision with B-26. Pilot Flyer 1c Suizu Mitsuo killed.	1 killed
3	10 Jul 42	Betty	No. 4 *Ku*	Crashed near the coastal village of Gaile some 35 miles south-east of Port Moresby, after being hit by AA fire over Port Moresby. *Hikotaicho* Lieutenant-Commander Tsusaki Naonobu and crew killed.	7? killed
4	11 Jul 42	Zero	Tainan *Ku*	Shot down by defensive fire from B-17s north of Papua. FPO3c Suzuki Matsumi killed.	1 killed
5	17 Jul 42	Rufe	Yokohama *Ku*	FPO1c Hori Tatsuo died of wounds sustained during combat or as a result of the crash landing after combat with a B-17 over Tulagi.	1 killed
6	18 Jul 42	Zero	Tainan *Ku*	Missing in bad weather during flight from Lae to Rabaul. Lt Kurihara Katsumi missing.	1 killed
7	18 Jul 42	Zero	Tainan *Ku*	Missing in bad weather during flight from Lae to Rabaul. FPO2c Miya Un'ichi missing.	1 killed
8	18 Jul 42	Zero	Tainan *Ku*	Missing in bad weather during flight from Lae to Rabaul. FPO1c Kobayashi Katsumi missing.	1 killed
9	18 Jul 42	Zero	Tainan *Ku*	Missing in bad weather during flight from Lae to Rabaul. FPO3c Onishi Yoshimi missing.	1 killed
10	23 Jul 42	Rufe	Yokohama *Ku*	Sea1c Matsui Saburo shot down and killed after combat with 11th BG Fortresses over Guadalcanal.	1 killed
11	1 Aug 42	Betty	No. 4 *Ku*	Flown and commanded by FPO1c Sekine Tokushiro, this Betty never returned from a night raid against Port Moresby and remains missing, almost certainly a victim of bad weather over New Britain.	7 killed
12	2 Aug 42	Zero	Tainan *Ku*	Flyer1c Motoyoshi Yoshio shot down and killed after combat with 28th BS Fortresses near Buna.	1 killed
13	2 Aug 42	Irving	Tainan *Ku*	Warrant Officer Tokunaga Tamotsu shot down by Airacobra after departing Lae for reconnaissance of Port Moresby.	1? killed
14	4 Aug 42	Rufe	Yokohama *Ku*	Sea1c Kobayashi Shigeto killed after being hit by defensive fire from B-17s and then crashed into one, off Guadalcanal.	1 killed
15	4 Aug 42	Babs	Tainan *Ku*	Shot down by RAAF Kittyhawk during reconnaissance of Milne Bay. Pilot FPO2c Hanahiro Keiryu and observer Warrant Officer Hasegawa Kameichi both killed.	2 killed
16	11 Aug 42	Zero	Tainan *Ku*	FPO3c Endo Masuaki wounded during combat with Kittyhawks over Milne Bay and ditched off Buna rather than risk returning to Rabaul.	
17	13 Aug 42	Zero	Tainan *Ku*	Lieutenant Murata Isao was lost with his fighter to unspecified operational causes at Lae, possibly a take-off or landing accident on Lae's badly damaged runway.	1 killed
18	14 Aug 42	Zero	Tainan *Ku*	FPO3c Arai Masami shot down by defensive fire from B-17s over Buna.	1 killed
19	17 Aug 42	Zero	Tainan *Ku*	FPO2c Norio Tokushige disappeared in bad weather during return to Rabaul after Port Moresby raid.	1 killed
20	23 Aug 42	Zero	Tainan *Ku*	Written off after landing accident at Buna. Pilot FPO2c Yamazaki Ichirobei OK.	

	DATE	TYPE	SERIAL/ UNIT	COMMENTS	FATALITIES*
21	26 Aug 42	Zero	No. 2 *Ku*	Shot down by Airacobras while taking off from Buna. FPO1c Iwase Ki'ichi killed.	1 killed
22	26 Aug 42	Zero	No. 2 *Ku*	Shot down by Airacobras while taking off from Buna. FPO3c Ihara Daizo killed.	1 killed
23	26 Aug 42	Zero	Tainan *Ku*	Shot down by Airacobras while taking off from Buna. FPO3c Nakano Kiyoshi killed.	1 killed
24	26 Aug 42	Zero	Q-102 / No. 2 *Ku*	Abandoned at Buna after being damaged during combat with Airacobras.	
25	27 Aug 42	Zero	Tainan *Ku*	Hit by small arms fire over Milne Bay and crashed. Pilot Lieutenant Yamashita Joji killed.	1 killed
26	27 Aug 42	Zero	V-130 / Tainan *Ku*	Hit by small arms fire over Milne Bay and ditched. Pilot FPO2c Kakimoto Enji captured.	1 POW
27	27 Aug 42	Zero	Tainan *Ku*	Shot down by Kittyhawks over Milne Bay. Pilot FPO1c Yamashita Sadao killed.	1 killed
28	27 Aug 42	Zero	Tainan *Ku*	Shot down by Kittyhawks over Milne Bay. Pilot Flyer1c Ninomiya killed.	1 killed
29	27 Aug 42	Val	No. 2 *Ku*	Shot down by Kittyhawks over Milne Bay. Pilot FPO2c Takahashi Koji and observer Lieutenant Yoshinaga Hiroshi killed.	2 killed
30	27 Aug 42	Val	No. 2 *Ku*	Ditched in Milne Bay after combat with Kittyhawks. Pilot Flyer2c Shibuya Masakichi killed, observer Flyer1c Koyamada Masami captured.	1 killed 1 POW
31	27 Aug 42	Zero	Tainan *Ku*	Shot down by Airacobras over Buna. FPO3c Matsuda Takeo killed.	1 killed
32	29 Aug 42	Betty	V-902 / Tainan *Ku*	Strafed and destroyed by Airacobras at Lae. Operated by Tainan *Ku* transport detachment.	
33	29 Aug 42	Betty	V-903 / Tainan *Ku*	Strafed and destroyed by Airacobras at Lae. Operated by Tainan *Ku* transport detachment.	
34	1 Sep 42	Val	Q-216 / No. 2 *Ku*	Landed on Papuan beach after running short of fuel following unsuccessful search for Allied ships at Milne Bay. Crew later killed in shoot out with Australian troops.	2 killed
35	1 Sep 42	Val	Q-218 / No. 2 *Ku*	Landed on Papuan beach after running short of fuel following unsuccessful search for Allied ships at Milne Bay. Crew later killed in shoot out with Australian troops.	2 killed
36	1 Sep 42	Val	Q-219 / No. 2 *Ku*	Landed on Papuan beach after running short of fuel following unsuccessful search for Allied ships at Milne Bay. Crew later killed in shoot out with Australian troops.	2 killed
37	7 Sep 42	Betty	Kisarazu *Ku*	Crashed landed on northern coast of Papua after raid against Port Moresby.	?
				* Total Fatalities (note other fatalities occurred in aircraft not classed as lost)	46

Breakdown of aircraft lost:

Zero	21
Betty	5
Val	5
Rufe	3
Nell	1
Irving	1
Babs	1
	<u>37</u>

APPENDIX 3 – CUMULATIVE AIRCRAFT LOSSES & FATALITIES
8 December 1941 to 8 September 1942

Allied Losses

	VOLUME 1 8 DEC 41 – 9 MAR 42	VOLUME 2 10 MAR – 30 APR 42	VOLUME 3 1 MAY – 18 JUN 42	VOLUME 4 19 JUNE – 8 SEPT 42	TOTALS
Airacobra	1	1	59	21	82
SBD Dauntless		1	37		38
F4F Wildcat	2	2	26		30
P-40E Kittyhawk / Warhawk		19	2	11	32
B-26 Marauder		8	12	9	29
B-25 Mitchell		4	15	4	23
A-24 Banshee		7	10	5	22
TBD-1 Devastator			15		15
Hudson	9	2	4	5	20
PBY Catalina	8	1	4		13
Wirraway	10	2		2	14
B-17 Flying Fortress	2	1	3	19	25
Beaufort				4	4
Lodestar				3	3
Other*	2	2	1	6	11
Totals	34	50	188	89	**361**
Fatalities	63	53	172	191	**479**

* Walrus, SOC Seagull (2), Ford Trimotor, F4 Lightning (2), Empire flying boat, DC-2. DC-5, Beaufighter, LB-30

Japanese Losses

	VOLUME 1 8 DEC 41 – 9 MAR 42	VOLUME 2 10 MAR – 30 APR 42	VOLUME 3 1 MAY – 18 JUN 42	VOLUME 4 19 JUNE – 8 SEPT 42	TOTALS
A6M2 Zero	2	26	41	21	90
G4M1 Betty	15	4	7	5	31
H6K4 Mavis	5		4		9
D3A1 Val	3		21	5	29
B5N Kate	2		35		37
E13A1 Jake	4		1		5
E8N2 Dave		3			3
F1M2 Pete			5		5
A5M4 Claude			6		6
A6M2-N Rufe				3	3
Unknown type			10		10
Other *		2	2	3	7
Totals	31	35	132	37	**235**
Fatalities	139	49	187	46	**421**

* H8K1 Emily, G3M2 Nell (2), E7K Alf, E9W Slim, J1N1-C Irving, Babs

SOURCES & ACKNOWLEDGMENTS

Research for this volume focuses on primary sources. As with all previous three Volumes, the private collections of Michael Claringbould and Peter Ingman contain considerable information obtained over many years for which it is not practicable to further credit, other than select sources listed below.

Special acknowledgements again go to Pacific War Air Historical Associates (PAWHA) members Ed DeKiep, Osamu Tagaya and Jim Lansdale (deceased) for their advice on Japanese aircraft markings; also to PAWHA member Luca Ruffato (deceased) for research on the Tainan *Ku* and translation of the *Kiyokawa Maru Sentoshoho*. Thanks also to Justin Taylan and his website www.pacificwrecks.com; to Gordon Birkett (Australia) for RAAF markings and to Steve Birdsall (Australia) for Coral Sea B-17 history. Thanks also to Russell Harada in Rabaul for translation work, and Axis History Forum contributor IJNFLEETADMIRAL (Mathew Jones). Also the ADF Serials and Combinedfleet websites.

BOOKS AND DOCUMENTS

Bullard, Steven (translator). *Japanese Army Operations in the South Pacific Area; New Britain and Papua Campaigns, 1942-43*. Australian War Memorial, Canberra, 2007.

Claringbould, Michael. *P-39/P-400 Airacobra vs A6M2/3 Zero-sen New Guinea 1942*. Osprey Publishing, 2018.

Claringbould, Michael and Ruffato, Luca. *Eagles of the Southern Sky*, Tainan Books, 2013.

Downs, Ian. *The New Guinea Volunteer Rifles NGVR 1939-1943 A History*. Pacific Press, Qld, 1999.

Feldt, Eric. *The Coastwatchers*. Penguin Group, Australia, 1991. First published 1946, plus Feldt's private letters collection in National Archives.

Gill, G. Hermon. Royal Australian Navy 1939-1942, Australia in the War of 1939-1945, Series Two Navy, Volume I. Canberra: Australian War Memorial, 1957.

Gillison, Douglas. *Royal Australian Air Force 1939-1942, Australia in the War of 1939-1945, Series Three Air, Volume I*. Canberra: Australian War Memorial, 1962.

Hickey, Larry. *Revenge of the Red Raiders*. IRP, 2006.

Lewis, Tom. *The Empire Strikes South*. Avonmore Books, Kent Town, SA, 2017.

Livingstone, Bob. *Under the Southern Cross The B-24 Liberator in the South Pacific*. Turner Publishing, USA, 1998.

McCarthy, Dudley. *South West Pacific Area First Year Kokoda to Wau*, Series One Army, Volume V. Canberra, Australian War Memorial, 1959.

Salecker, Gene Eric. *Fortress Against the Sun The B-17 Flying Fortress in the Pacific*. USA: Da Capo Press, 2001.

Tagaya, Osamu, *Mitsubishi Type 1 Rikko Betty Units of WW2*, Osprey Publishing 2001

Vincent, David. *The RAAF Hudson Story Book 2*. Vincent Aviation Publications, Highbury SA, 2010.

AUSTRALIAN WAR MEMORIAL
File Series 52

War History/ Unit and Commander War diaries/Divisions and Force/ NG Force HQ 1/5/51/10; 521/68/ Thursday Island War Diary 7/8/10; 2-1Light AA 4/4/1; 23AA 4/16/26.

File Series 54

779/3/14, 779/3/75, 779/3/99, 779/3/25 1944-45, 779/3/3 POM and Milne Bay #2-6, 779/3/98 NZ reports mostly Guadalcanal, 779//7-1 1944/45 captures New Guinea, 779//7-2 1944/45 captures New Guinea, 115/13/2, 423/4/73, 423/11/82, 576/2/5, 812/3/17,

File Series 64

1/102 handwritten diary 11/20 Squadron, 1/105, 1/103 RAAF unit history sheets forms 50 & 5, AWM A9186 series

Australian Military Forces, official history of No 8 Military District, Port Moresby (June/ July 1942)

AUSTRALIAN AND US NATIONAL ARCHIVES

B-17 missions log, Battle of Coral Sea, Parts I and II

Naval Staff history, Battle Summaries 45 and 46 Battles of Coral Sea & Midway 1952

Feldt, Eric, private letters pertaining to Port Moresby and Tulagi ops, NAA B3476, 49C

RAAF patrol logs, numerous

OTHER SOURCES

Lt-Col John Fields, USAAC B-17E pilot, private diary of, published by Kenneth Fields 9 October 1982

RAAF 6, 11, 20, 24, 32, 75 & 76 Squadron operations logs July to September 1942.

Microfilm histories/ documents pertaining to 5th Air Force establishment, 8th Fighter Group, 35th Fighter Group, 435th Reconnaissance Squadron, 22nd Bombardment Group, 3rd BG including 90th BS, 13th BS, 8th BS.

Morison, Coral Sea, Midway and Submarine Actions, May 1942 – August 1942

Miscellaneous private letters, diaries, microfilms, captured documents, in possession of authors.

JAPANESE LANGUAGE SOURCES

Admiral Kusaka Junichi, private diary hand-written in kanji

Organization of the Naval Air Groups: Wartime Organization, 15 November 1940, Secret Order No.824. Revised 1941, No.366, No.618, No.994, No.1171, No.1497.

Senshi Sosho Volume 49

Iwashige Tashiro, The visual guide of Japanese wartime merchant marine, KK, Tokyo, 2009

Fukui Shizuo, Kaigun Kantei shi, 1869-1945: *Kokubokan, Suijokibokan, Suirai* to *Sensuibokan*, KK *Besutoserāzu*, Tokyo, 1982

Mori Tsunehide, Gunkan Zakki Cho, Tamiya, Shizuoka

Nohara Shigeru, "*Nihon Teikoku Kaigun Suijo Teisatsuki*", FAoW no 47, Bunrin-do KK, Tokyo

Nemoto Kumesako, Warrant Officer, diary, pilot with seaplane tender *Kiyokawa Maru* Oct 1941 to Jun 1942 (translated AWM).

Japanese Naval Units and Air Bases - ATIS captured document dated 26 Nov 42 (via Jim Lansdale)

Organization & history of 25 Air Flotilla, ATIS captured document dated 16 Oct 42 (via Lansdale)

Manual of Military Secret Orders ATIS captured document dated 20 Jul 43 (via Lansdale)

Japanese Naval Air Service Intelligence memorandum No 28, ATIS captured document dated 7 Jul 43 (via Lansdale)

Japanese Air Terms, Squadron Leader A. R Boyce, Far Eastern Bureau, Calcutta 1944

Japanese naval Air Organization CINCPAC Bulletin Nos. 16 -45

The Achilles Tendon of the IJN by Yoshida Toshio (CDR, 59th class, Secretary to ADMs Yonai, Shimada, and Nagano. Member of the Naval General Staff & Imperial General HQ)

BBKS Kan-49, p. 49, NAII RG457 Translation of Japanese Navy Messages, Japanese Naval Forces, 26 Mar 1942 - 16 April 42

Translation of Japanese Navy Messages, Japanese Naval Forces, 17 April 42 –30 April 1942

Suiraisentai Senjinisshi 17.1.31-1 (floatplanes used by 4th Fleet)

Japanese Monographs, US Library of Congress, post-war summaries conducted by Historical Section G2, volumes pertaining to South Seas operations

Chubu Taiheyo homen Kaigun sakusen, Kan-38, *Asagumo Shimbunsha*, Tokyo, 1970.

Register of requisitioned vessels, Mobilization Bureau, Navy Ministry, 1 June 1944

Tabulated Records of Movement (TROMs) for relevant Japanese ships cited in text.

Yasuo Izawa, The Land Attack Force, Toyko 1995 ISBN 4-257-17299-1

Other papers/ documents/ diaries in Japanese language sourced via Japanese National Institute for Defence Studies.

Interrogations of senior IJN officers 17 October 1947 in Tokyo by USN; Lieutenant-Commander Sekino Hideo of No 6 Cruiser *Sentai*, Captain Yamada Morishige, Operations Officer of 5th Air Flotilla, stationed on *Zuikaku*, Captain Kijima Kikunori, Chief of Staff 6 *Sentai*, aboard *Aoba* (held Australian Archives)

Japanese Newspapers

Japan Times and Advertiser, newspaper, articles June to September 1942

Asahi Shimbun, newspaper, articles June to September 1942

Unit Operations Logs (*Kodochosho*)

Chitose *Kokutai*, No. 4 *Kokutai*, No. 6 *Kokutai*, Kisarazu *Kokutai*, No. 17 *Kokutai*, Genzan *Kokutai*, Yokohama *Kokutai*, No. 14 *Kokutai*, Tainan *Kokutai*, Misawa *Kokutai* & Chitose *Kokutai*.

Ship Operations Logs (*Sentochosho*)

BBKS, *Kiyokawa Maru Sentoshoho*, BBKS, No. 6 *Sentai Sentoshoho* (various cruisers and ships)

Photo Credits

Author's collections, AWM, and Ed DeKiep collections

INDEX